DATE			

And, as his strength
Failed him at length,
He met a pilgrim shadow—
"Shadow," said he,
"Where can it be—
This Area 51?"

"Over the mountains
Of the Moon,
Down the Valley of the Shadow,
Ride, boldly ride,"
The shade replied—
"If you seek for Area 51!"

(With apologies to Edgar Allen Poe's poem "Eldorado,"
also about a place which officially did not exist.)

AREA 51
BLACK JETS

BILL YENNE

Zᴇɴɪᴛʜ Pʀᴇss

First published in 2014 by Zenith Press, a member of Quayside Publishing Group, 400 First Avenue North, Suite 400, Minneapolis, MN 55401 USA

© 2014 Zenith Press
Text © 2014 Bill Yenne

Zenith Press titles are also available at discounts in bulk quantity for industrial or sales-promotional use. For details write to Special Sales Manager at Quayside Publishing Group, 400 First Avenue North, Suite 400, Minneapolis, MN 55401 USA.

To find out more about our books, visit us online at www.zenithpress.com.

ISBN-13: 978-0-7603-4426-2

Frontis: Pointing the way in the Nevada desert. Turn right on Groom Lake Road at the next intersection. *Bill Yenne*

Title page: As strange and mysterious as it appears in this image, the Bird of Prey was a real airplane that really flew—and flew in complete secrecy—in the skies above the place that had come to be known as Area 51. *Illustration by Erik Simonsen*

Contents: Looking futuristic in 1964, the Lockheed YF-12A still looks futuristic today. A product of the golden age of the renowned Skunk Works, it was one of the great signature airplanes of the place called Area 51.

Library of Congress Cataloging-in-Publication Data

Yenne, Bill, 1949-
    Area 51-- black jets : a history of the aircraft developed at Groom Lake, America's secret aviation base / Bill Yenne.
            pages cm
Summary: "An illustrated history of the U.S. military's development of secret spy aircraft at Groom Lake, a.k.a. Area 51"-- Provided by publisher.
    ISBN 978-0-7603-4426-2 (hardback)
    1. Area 51 (Nev.)--History. 2. Aeronautics, Military--Research--United States--History. 3. Research aircraft--United States--History. 4. Airplanes, Military--United States--History. 5. Area 51 (Nev.)--Pictorial works. I. Title. II. Title: Black jets, a history of the aircraft developed at Groom Lake, America's secret aviation base.
    UG634.5.A74Y46 2014
    623.74'6--dc23
            2013033111

Editorial Director: Erik Gilg
Editor: Caitlin Fultz
Design Manager: James Kegley
Cover Designer: Matthew Simmons
Layout Designer: Diana Boger

Printed in China

10 9 8 7 6 5 4 3 2 1

CONTENTS

PROLOGUE

WHAT *IS* AREA 51?

IN THE CARTOGRAPHY of our lives, our dreams, and our popular culture, certain landmarks are embodied with great meaning that transcends their identity as mere places. We travel to certain ones to see things of great cultural or historical importance. Whether we are fond of Michelangelo or Picasso, or of Winslow Homer or Warhol, we go to great museums to view and marvel at celebrated works of art. We go to the Smithsonian to see the tangible artifacts of American history, from the Star Spangled Banner to the Spirit of St. Louis.

Meanwhile we visit other iconic places, such as Waikiki or Las Vegas or any numbers of Disney Worlds or Lands or their analogues with other themes, for specific genres of "fun."

And finally there are places we visit not for specific artifacts or specific amusements but for the intangible reason that we just want to be there or, arguably more importantly, to say that we have been there.

We go to such places to breathe a certain rarified air.

We go to such places—Times Square or the corner of Haight and Ashbury—not so much to see and touch specific things, but to stand there and sense an ethereal yet palpable energy or to soak up the vibe.

This genre of venues possesses an importance that is greater than the sum of its parts. Merely the mention of one word, such as "Sturgis" or "Graceland," speaks volumes to those who venerate these places for what they represent. Of course none of these places is for everyone, and that is what makes each of them so important and so special to those for whom they do resonate. For such people, even those who have never been to these places that are the nexus of their fascination, the mere mention of the name is like a mantra that is a key to unlock an emotion.

Area 51 is such a place.

Even though Area 51 devotees cannot actually go there and stand in the gravel and fine desert dust of its epicenter, they come by the thousands every year to look at the mountains beyond which lies their field of dreams. The fact that armed guards prevent them from completing their pilgrimage only adds to the rush of excitement and the belief that this place is truly special.

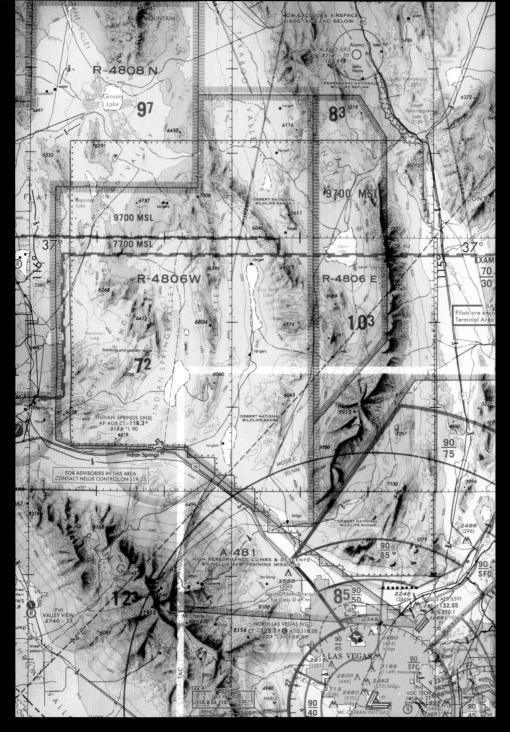

This FAA aeronautical chart shows the proximity of Las Vegas to the place known as Area 51. The box at top left labeled as "R-4808N" with Groom Lake in the center is the restricted air space known to Air Force pilots as Dreamland. The large airfield complex at Groom Lake, originally known as Watertown and built by Lockheed and the federal government in the late 1950s, officially does not exist and therefore appears on no US government maps. By some unofficial calculations, Area 51 comprises this whole box, by others, it is a smaller box centering on Groom Lake. *FAA*

In the 1980s, when the term "Area 51" entered the lexicon of popular culture, it was one of those legendary but not quite mythic places like Atlantis or Shangri-La or El Dorado. There are those who vociferously believe that these places truly exist while readily admitting that their precise locations are a bit vague. In turn the ambiguity and obscurity add to an excitement that is kept alive by the conspiracy theories woven by and for the true believers.

What is especially curious and paradoxical about this forbidden land called Area 51 is that it lies so close to Las Vegas, Nevada. In the cartography of our lives, our dreams, and

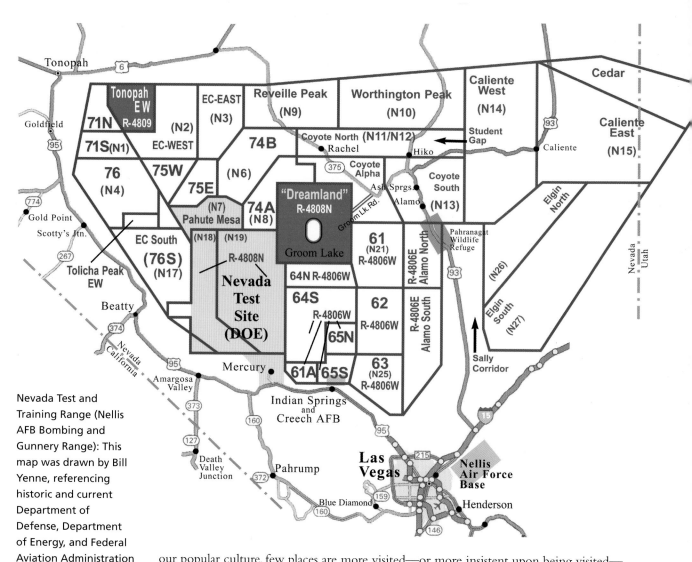

our popular culture, few places are more visited—or more insistent upon being visited—than Las Vegas.

The man who did more than any other to catapult the idea of an area numbered 51 into the public consciousness was a self-described engineer named Bob Lazar. In the late 1980s, he went to the media, specifically to George Knapp at KLAS television in Las Vegas, with his vivid stories of having worked at Sector 4 (S-4) of Area 51, where he had seen evidence that the US government possessed and was studying otherworldly spacecraft and their extraterrestrial crewmen. There was even a suggestion that the US Air Force was testing the alien spacecraft in their desert hideout.

Lazar was taken seriously—even welcomed and embraced—by those who had long believed not only in the existence of such craft and green or gray beings, but also in a thoroughly institutionalized government cover-up dating back to at least the 1950s. Lazar's detractors, meanwhile, cite a lack of concrete evidence and the fact that no record of him exists at his claimed alma maters, the Massachusetts Institute of Technology (MIT) and the California Institute of Technology (Caltech).

Nevertheless Lazar helped create and shape a popular mythology that spawned numerous books, movies, and magazine articles, especially during the 1990s. For instance, the idea of a government conspiracy that keeps extraterrestrial beings secret from the

Above: When Lockheed engineers first came to test airplanes at Area 51, the striking appearance of the Archangel/Oxcart/Blackbird family was not yet imagined and would have seemed strange and otherworldly. After half a century, it *still* seems that way. *Lockheed*

Left: Groom Lake is a dry lake most of the time, but a heavy rain can change that. This photo was taken across the lake from the hangar complex on April 6, 1965. The fish is probably another "Area 51 hoax." *Tony Landis collection*

public was the catalyst for the popular television series *X-Files*, which aired for nine seasons beginning in 1993. The Area 51 conspiracy theory also became a staple of Art Bell's popular late-night radio show *Coast to Coast AM*, which reached fifteen million listeners during its heyday in the late 1990s. Then too, Las Vegas's minor league baseball team, the Las Vegas Stars, renamed itself as the Las Vegas 51s and adopted a cartoon extraterrestrial as their mascot and logo.

Whatever one may say about Bob Lazar and his legacy, it can certainly be said that he played the key role in putting Area 51 on the map. Having said that, one must follow up with the question of exactly *where* on the map it is.

Lazar was right that Area 51 really exists, and that it is at a secret, previously undisclosed location within the more than 4,680 square miles of desert, mountains, and restricted airspace across parts of Clark, Lincoln, and Nye counties that comprise the Nevada Test and Training Range (NTTR). Within this airspace, larger that the state of Connecticut, the US Air Force has long conducted training operations such as the well-known Red Flag combat simulation exercises. Also known (and so identified on perimeter signage) as the Nellis Bombing and Gunnery Range, the NTTR is controlled from Nellis AFB, which is located eight miles northeast of Las Vegas. Through the years, the NTTR has grown—and continues to grow—through the acquisition of adjacent public land formerly administered

This is perhaps the first aerial photograph of the Groom Lake complex ever released by the US Air Force. Groom Lake is directly above the F-22 in this photograph taken by Senior Airman Matthew Lancaster on March 5, 2013, during the Red Flag 13-3 exercise over the Nellis AFB Range. By the time of this unintended photo release, the existence of the "nonexistent" complex was so well known that denial had faded to a mere refusal to acknowledge it, but even this was impractical. The Internet abounded with unofficial telephoto lens images, and anyone could "tour" the base on Google Earth. A little more than three months later, in June 2013, the CIA released a 1992 document containing a map showing that the airfield complex at Groom Lake had always been known officially as "Area 51." *USAF*

by the Department of the Interior. It also surrounds, on three sides, another off-limits world: the 1,360 square miles of the Department of Energy's Nevada National Security Site (N2S2). Previously known (and referred to in this book) as the Nevada Test Site (NTS), this is where nuclear weapons were tested for more than four decades.

It has been well known for many years that the US Air Force, as well as the CIA, conducted flight-test programs for secret airplanes at an air field at Groom Lake, a dry lakebed within the NTTR that is roughly 84 miles northwest of Las Vegas. The fact that it went for so long without official acknowledgement lent an aura of mystery to the open secret. As late as 1984, *Aviation Week & Space Technology*, the industry journal that has a reputation for the best inside information about the aviation world, mistakenly referred to Groom Lake as "Broom" Lake.

Through the Cold War, the US Air Force gradually imposed more restrictions on the airspace over the entire range, and by the 1960s, even air force pilots were being warned not to enter the 600 square miles above and around Groom Lake. To air traffic controllers and air force personnel, the mysterious place soon came to be known informally as "Dreamland." Federal Aviation Administration (FAA) charts place Groom Lake within a swath of restricted airspace called R-4808N.

In an August 13, 1984, article in *Aviation Week*, it was referred to furtively as "the north secret test range of Nellis Air Force Range." At that time, as the magazine reported, the US Air Force had just closed 90,000 acres of public land in the Groom Mountains contiguous to the eastern boundary of "the north secret test range" without filing an environmental impact statement. John O. Rittenhouse, deputy for installations management to the deputy assistant secretary of the air force for installations, environment, and safety, told the public lands and national parks subcommittee of the House of Representatives Interior and Insular Affairs that top-level Reagan administration officials made the decision to close the range to public access for "national security reasons."

Aviation Week added that Rittenhouse had explained that the area was to be used "only as a buffer and not for . . . 'environmentally significant activities.'"

In other words, it was attached to the range to keep out prying eyes who might be tempted to turn their binoculars upon the secret world of the air field at Groom Lake.

In the 1950s, long before it had taken on its cloak of extraterrestrial mystery, the air field was known as the Watertown landing strip—more than a bit fanciful, given that it was located on Groom's lakebed, which is almost perpetually bone dry. Later in that decade, the foundation for later conspiracy theories was laid when Watertown became a rookery for

CIA spyplanes, a secret base where they could rest between overseas deployments. These operations are discussed in detail in the CIA's official history of the Aquatone and Oxcart spyplane programs, written by Gregory Pedlow and Donald Welzenbach and published in 1992. Copies of this report, albeit in redacted form, have been widely circulated among aerospace historians since the turn of the century.

Throughout the past few decades, most people in the aerospace world have referred to the air field complex as "Groom Lake," just as those in the pop culture world have called it Area 51.

Though the Groom Lake complex was known to exist, and Dreamland well known as restricted airspace, inquiries to any government agency—military or civilian—always elicited the uniform response that there was no such place. You might as well be asking a stranger for directions to the Emerald City of Oz.

One of the strangest aspects of the CIA and air force denial of their "Emerald City" was their reticence to acknowledge that they had used the term Area 51 themselves. One of the few potentially real official mentions came in a May 1967 memo appearing to be from Director of Central Intelligence Richard Helms. This document, in which aircraft deployments from Area 51 are discussed, has been in circulation for a number of years, but it has not been authenticated.

The first official acknowledgement that the place was *officially* called Area 51 came on June 25, 2013, with the release of a slightly less redacted version of the 1992 Pedlow and

In its aerial photography of much of the United States, the US Geologic Survey took this photo of the Groom Lake complex in the 1970s. *Tony Landis collection*

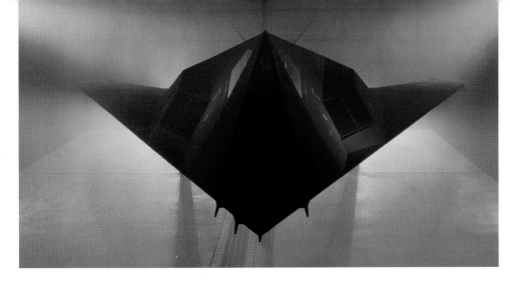

Welzenbach report. In it we now see that the authors refer to the Groom Lake base as Area 51 on sixteen pages, although the index shows that at least three additional references to Area 51 remained redacted.

It turns out that Pedlow and Welzenbach had also officially put Area 51 on the map at about the same time that Bob Lazar was doing so unofficially. In the 2013 release, a previously redacted map places Area 51 within Area R–4808N as seen in the FAA map on page seven in this book and within the red box at the center of our map on page eight.

Why did the CIA redact the casual references to Area 51 by Pedlow and Welzenbach when their very detailed report was first released many years ago? We don't know and perhaps we never will. This only serves to add to the mystery surrounding Area 51, but perhaps this was not unintentional. By allowing the term to be the sole property of the UFO speculators for a quarter of a century, the CIA and the air force may have hoped for a diversion of serious scrutiny.

However, despite the redactions of the term in the 1992 report, it was in this same time frame that Area 51 was becoming the focus of a great deal of attention and speculation, some of it serious, much of it on the strange side. In the 1980s, and especially in the 1990s after Bob Lazar's stories became part of extraterrestrial folklore, a growing number of people began bringing their binoculars to the desert, where the well–guarded perimeter of Area 51 only gave rise to suspicions that the conspiracy theories bore at least a modicum of substance.

Both extraterrestrial believers and the merely curious came to visit, hoping to catch a glimpse of mysterious lights in the sky and to marvel at the mysterious "black mailbox," where it was once theorized that the extraterrestrials and/or the keepers of those extraterrestrials picked up their mail. Painted white since the late 1990s, the mailbox actually belongs to rancher Steve Medlin.

In 1996, as the interest in Area 51 promised to lure even more tourists to this remote corner of the state, the state of Nevada even went so far as to designate its State Highway 375 (formerly State Highway 25) as the "Extraterrestrial Highway."

As the initial excitement about the extraterrestrial theory faded, those who had come to drive this highway discovered that there *were* strange, unidentifiable lights in the sky. They were *literally* seeing flying objects that they could not identify! There really *were* UFOs in the skies over Area 51, but they were not of another world; they were from ours.

And so it was that the "black airplane" cult grew up upon ground fertilized by theories of the unexplained that had their origins among those who have longed for the extraterrestrials to be real.

The black airplanes, though, *are* real, and this is their story—or at least as much of it as we can tell.

CHAPTER 1
HOW NOWHERE BECAME SOMEWHERE

THE AREA AROUND the Nevada Test and Training Range (NTTR) and the Nevada National Security Site (N2S2) is about the closest approximation to the idea of "the middle of nowhere" that can be found anywhere in the contiguous United States. One can easily drive for an hour with but scant evidence of human habitation and for a half hour on Highway 375 without passing another car.

For centuries, what is now southern Nevada, and indeed most of the state, remained largely empty because of its harsh climate and widespread lack of water. Only small numbers of Paiute and Shoshone lived here when Anglo Americans first passed through what are now the NTTR and N2S2 in the nineteenth century, and they lived only in the relatively few fertile valleys. Early in that century, westward-bound wagon trains generally avoided southern Nevada. Later, inspired by the Comstock silver boom in northern Nevada, prospectors established a scattering of mainly disappointing mines.

The abandoned homes of nineteenth century ranchers and prospectors can still be found in remote corners of the Nevada Test Site and the adjacent Nellis Range. *DOE*

Military personnel casually observe a 22-kiloton blast at Frenchman Flat in Area 5 of the Nevada Test Site on April 15, 1955. *NNSA/NSO*

By the twentieth century, there were a few ranches and a few small towns, such as Tonopah, Goldfield, and Rhyolite. To the south, Las Vegas was incorporated in 1911. While Las Vegas flourished as a railroad town and later as a gaming mecca, places like Rhyolite withered into ghost towns as the mines became unsustainable.

In October 1940, President Franklin Roosevelt signed off on the idea of turning more than three million acres of government-owned land in the "middle of nowhere" north of Las Vegas over to the US Army Air Corps as the Las Vegas Bombing and Gunnery Range. The Las Vegas Army Airfield (now Nellis AFB) was established on the north side of Las Vegas in 1941 to manage operations over the range, which included the Army Gunnery School, started in 1942.

Additional auxiliary fields managed by the main base were set up at places such as Indian Springs, near Las Vegas on the southwest corner of the range, as well as at Pahute Mesa, Groom Lake, and near Tonopah. Las Vegas AAF was renamed as Nellis AFB in 1950, while Indian Springs became Creech AFB in 2005 and is now well known as the home to MQ-1 Predator and MQ-9 Reaper unmanned aerial vehicle operations. Tonopah is active, but mainly in secrecy, while Groom Lake officially does not exist.

The 24-kiloton "Nancy" test, seen here, was conducted on March 24, 1953, as part of Operation Upshot-Knothole. *NNSA/NSO*

Dr. Alvin Graves, scientific advisor for the nuclear tests at the Nevada Test Site, points out a location near Beatty, Nevada, for support director Seth Woodruff and test manager James Reeves in March 1955 during Operation Teapot. Graves served at Los Alamos during the Manhattan Project. *NNSA/NSO*

After World War II, another slice of the huge range became famous for another purpose. Even as the future Nellis AFB was finding its place in the postwar air force, the United States was assembling a growing arsenal of nuclear weapons. Managing this stockpile was the responsibility not only of the armed forces, but also of the new Atomic Energy Commission (AEC), which was set up in 1946 to oversee American nuclear programs.

These weapons required testing, which naturally required a place for the tests. The first test had occurred in New Mexico in July 1945, but the early postwar tests took place at

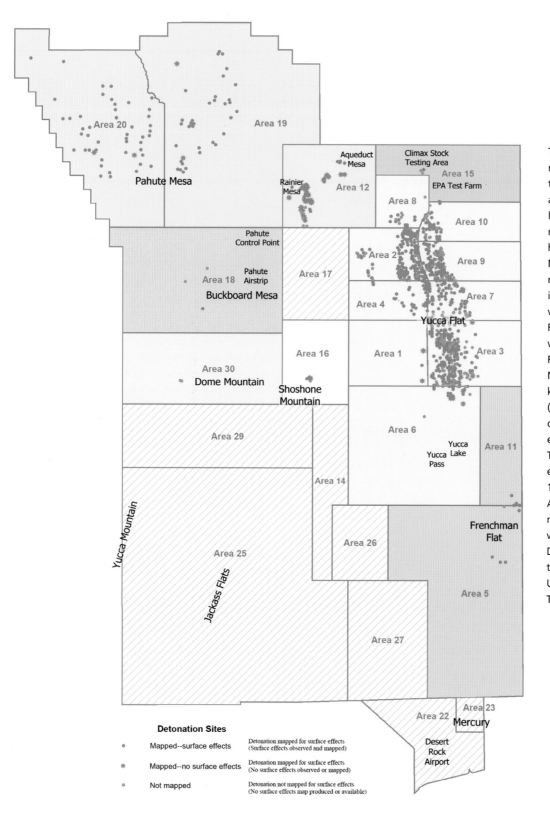

This map shows the numbered areas of the Nevada Test Site, as well as the specific locations of underground nuclear tests conducted here through 1992. Most above-ground nuclear tests took place in Areas 1 through 8, which include Yucca Flat, and in Area 25, which includes Jackass Flats. The segment of Nellis Range air space known as Dreamland (encompassing Area 51), overlays the northeast edge of the Nevada Test Site, including the eastern edges of Areas 15 and 10. The Pahute Airstrip in Area 18 remains active. This map was prepared in 2003 by Dennis Grasso as part of the Geologic Effects of Underground Nuclear Testing project. *USGS*

Detonation Sites

- Mapped--surface effects — Detonation mapped for surface effects (Surface effects observed and mapped)
- ∗ Mapped--no surface effects — Detonation mapped for surface effects (No surface effects observed or mapped)
- Not mapped — Detonation not mapped for surface effects (No surface effects map produced or available)

remote islands in the Pacific Ocean for security reasons. However, the complex logistics required for the Pacific tests led the AEC and the services to move the testing to the contiguous United States. President Harry Truman initially disagreed with the AEC and supported a continuation of the Pacific tests, but against the backdrop of the Korean War, he reconsidered, and in October 1950, he gave the green light.

In 1947, the search for a site and the development of nuclear weapons doctrine was undertaken under the Armed Forces Special Weapons Project (AFSWP, pronounced AF-SWAP), codenamed Project Nutmeg.

According to Barton Hacker in his book *Elements of Controversy*, AFSWP Search Committee Director Howard Hutchinson was asked to assess the "physical feasibility [and the] how, when, and where" of conducting nuclear weapons tests in the continental United States without, and as Hutchinson himself put it, radioactive fallout causing "physical or economic detriment to the population." Hutchinson concluded that "properly engineered sites, under proper meteorological conditions [such tests would] result in no harm to population, economy or industry."

Remote locations across the desert Southwest were considered, but so too were areas along the coasts of Maine, Maryland, Delaware, Virginia, and the Carolinas. The most ideal of these was thought to be the area north of Cape Fear, an arguably appropriate name for a nuclear test site.

By this time, the Nutmeg team had essentially picked the best possible continental site. They had come to the Las Vegas Bombing and Gunnery Range, experienced its vast emptiness, and realized that this "middle of nowhere" would be a good place for the "somewhere" of American atmospheric nuclear testing. Title to the 1,360-square-mile slice of the bombing range was formally transferred from the US Air Force to the AEC, though the air force controlled the air space and dropped many of the early nuclear weapons.

The Atomic Energy Commission "company town" of Mercury, Nevada, was founded in 1950 within the Nevada Test Site as the spartan facility known as Base Camp Mercury. Within a year, a $6.7-million construction project was transforming it into the sprawling city seen here. *NNSA/NSO*

Initially known as the Nevada Proving Grounds, and later as the Nevada Test Site (NTS), it was established due north of Indian Springs on January 11, 1951. The first nuclear test here took place as part of Operation Ranger just sixteen days later.

Initial uneasiness about the tests in Las Vegas gradually dissipated as the AEC undertook a massive construction program, pumping many federal dollars into a local economy that had yet to reinvent itself as a destination resort. The building boom included the new "company town" of Mercury, Nevada, 65 miles northwest of Las Vegas, which housed AEC staff. It eventually had a population of around 10,000. Indeed, Las Vegas even embraced the testing, with people gathering on rooftops for parties organized around the viewing of mushroom clouds rising from the distant test site.

Guards Milton Miller and John Metcalf inspect the pass of a nervous Frank Waters at the entrance to the Nevada Test Site in 1953 (when it was still called the Nevada Proving Grounds). Waters worked for the Joint AEC-DOD Test Information Office. Miller and Metcalf were employed by Federal Services, Inc., a contractor that supplied security services. *NNSA/NSO*

During the years of extensive atmospheric testing at the NTS, little was done to conceal the program from the general public. Indeed, with mushroom clouds visible from Las Vegas, little *could* be done. However, many aspects of the program remained under wraps, many of them remaining secret through the end of the century and beyond.

One intriguing clandestine aspect of NTS operations during the atmospheric testing era involved the Lookout Mountain Laboratory, which remained classified for over a quarter century after it closed and is hardly spoken of even today. Indeed, Lookout Mountain barely registers on the radar of the Area 51 conspiracy theorists who passionately seek out shadowy government and contractor entities.

Amazingly, the Lookout Mountain Laboratory existed right under the noses of the national media in the Hollywood Hills! The laboratory was actually a state of the art motion picture studio that was secretly established in 1947 in a 100,000-square-foot

A film crew from the Lookout Mountain Laboratory records Wasp Prime, the 3.3-kiloton explosion of an air-dropped weapon that occurred on March 29, 1955, in Area 7. Lookout Mountain was a bigger secret than much of what happened at the test site. *NNSA/NSO*

building on Wonderland Avenue in Los Angeles. The facility was originally built in 1941 as an air defense radar coordination center. From 1947 to 1969 it operated secretly as the Lookout Mountain AFS, providing production services for classified motion pictures and still photography for the Department of Defense (DOD) and the AEC at the NTS, at the nuclear and missile test sites in the Pacific, and elsewhere.

Crews assigned to the Lookout Mountain field staff included photographers who were airmen assigned to the Air Force 1352nd Photographic Squadron and shot footage of nuclear tests in formats including CinemaScope, VistaVision, and 3D. Back at Wonderland Avenue, the studio processed 35mm and 16mm film, as well as still photographs. Civilians who worked at major studios such as Warner Brothers, MGM, and RKO also worked clandestine side jobs as cameramen and directors at and for Lookout Mountain. The site is now a private residence.

Through 1992, when they officially ended, 928 announced nuclear tests involving 1,021 detonations were conducted at the NTS. Of these, 90 percent were underground tests, which were conducted in accordance with the Nuclear Test Ban Treaty of 1963.

At the same time that the AEC was conducting weapons–related tests at the NTS, the agency was also looking ahead to nuclear propulsion for a wide range of vehicles, including aircraft and missiles. The former were evaluated under the Nuclear Energy for Propulsion of Aircraft (NEPA) program, the latter under Project Pluto. Ultimately nuclear reactors were adapted as propulsion systems for ships and submarines, but the weight of the reactors and necessary shielding made it impractical for airplanes and other smaller vehicles.

The cab of the 700-foot tower from which the 44-kiloton Smoky device was detonated in August 1957. At the time, the tower was the tallest structure in Nevada. In left foreground are shelters being tested for the French and German governments. On right are military vehicles to be used to evaluate blast effects. The trench area from which military observers view the nuclear test is 4,500 yards from the tower. The troop assembly area was 9,800 yards away, and News Nob, from which the media had watched the tests since 1952, was 17 miles away. *NNSA/NSO*

Part of Operation Plumbbob, the 37-kiloton Priscilla weapon was detonated on June 24, 1957, as it was held aloft aboard a balloon. *NNSA/NSO*

Another application that seemed practical at the time was space flight. In 1955, the AEC and the US Air Force initiated an effort to develop a space launch vehicle capable of taking people to Mars. This project, Project Rover, led to an experimental low-power test firing of a nuclear rocket engine, designated Kiwi A, in July 1959. In 1960, Rover evolved into the Nuclear Engine for Rocket Vehicle Application (NERVA) program, which was tested at the Nuclear Rocket Development Station within the NTS at Jackass Flats, sharing some of the Project Pluto infrastructure.

A second test, designated as Kiwi A-Prime, was conducted at full power in July 1960. The first of a series of Kiwi B tests was conducted in December 1961. This series was followed by an unfolding series of reactor and engine experiments with such colorful code names as Phoebus, Peewee, and Nuclear Furnace. Meanwhile the parallel Reactor In-Flight Test (RIFT) program was also intertwined with NERVA. Nearly two dozen reactor and rocket engine tests, including those of the Experimental Engine Cold Flow (XECF) experimental nuclear rocket engine program, were conducted at the Nuclear Rocket Development Station through 1971, with some engines operating for as long as an hour and a half.

In early 1969, National Aeronautics and Space Administration (NASA) officially unveiled its plan to send a six-man crew to the Martian surface. As described in official 1969 NASA documents, six astronauts would rendezvous with the NERVA-powered Mars vehicle in Earth orbit and depart for Mars on November 12, 1981. They would arrive at Mars on August 9, 1982, and leave the NERVA ship in orbit while they descended to the Martian surface for a ten-week visit. The astronauts would then depart Mars on October 28, 1982, conduct a close flyby of Venus on February 28, 1983, and return to Earth on August 14, 1983.

However, in August 1969, shortly after Apollo 11's first human landing on the moon, the Nixon administration quietly cancelled the 1981–1983 Mars mission. NERVA continued until 1973, when it, too, was officially cancelled.

Meanwhile, as operations at the NTS were evolving, neighboring Nellis AFB was a growing Air Training Command (ATC) fighter training base. The US Air Force sought to greatly increase its number of fighter pilots, and the ideal flying weather in southern Nevada made it a good venue. As Las Vegas Airfield, and as Las Vegas AFB after 1948, the base had functioned as a subsidiary of Mather AFB near Sacramento and later of Williams AFB in Arizona. In October 1950, this increasingly important base was upgraded to primary installation status, and all the base management functions were transferred from Williams.

It was during this time that ATC started its Aircraft Gunnery School at the base and activated the 3525th Aircraft Gunnery Squadron, which utilized the services of World War II air combat veterans to train aircraft gunnery instructors. Meanwhile, under the umbrella of the 3595th Pilot Training Wing, Nellis became a key location for transition training of pilots experienced in piston-engine fighters during World War II to jet fighters from 1950, though the end of the Korean War. Three years later, Nellis was a center of combat crew training for fighter pilots, mainly in F-86 Sabres.

On the first day of 1954, as the mission of the base reverted to gunnery training, the Aircraft Gunnery School was redesignated as the US Air Force Fighter Weapons School. The school still carries on, although it had undergone a series of redesignations and has been complemented by a nonflying Fighter Weapons Center since 1966. The center became the US Air Force Warfare Center in 2005.

When the F-100 Super Sabre was introduced into service as a successor to the F-86, ATC requested the aircraft for its Fighter Weapons School, but the aircraft were delivered only to operational units of the Tactical Air Command (TAC). In January 1957, according to Thomas Manning in his book *History of Air Education and Training Command, 1942–2002*, ATC commander General Charles Meyers told Air Force Chief of Staff General Nathan Twining that "the only way ATC could continue to operate the school was if the Air Force would agree to provide first-line aircraft on a timely basis." If that couldn't be agreed upon, he felt that the Fighter Weapons School should be transferred to TAC.

Twining called his bluff, and the transfer was made in February 1958. Nellis AFB itself was transferred to TAC five months later.

Right: The NERVA reactor is ready to be hooked up at Test Cell C within the Nuclear Rocket Development Station in Jackass Flats. *NNSA/NSO*

Below left: On December 8, 1962, President John F. Kennedy came to visit the Nuclear Rocket Development Station. Having committed the United States to land a man on the moon, Kennedy was anxious to see an American flag also flying on Mars. *NNSA/NSO*

Below right: The Kiwi program was part of Project Rover, an early program aimed at developing a nuclear reactor for space travel. Kiwi-A Prime, seen here, underwent a successful full-power run at Jackass Flats on July 8, 1960. *NNSA/NSO*

Late on December 1, 1967, this nuclear rocket engine was transported to the Nuclear Rocket Development Station in Jackass Flats. The first ground experimental nuclear rocket engine (XE) assembly (left) is shown here in "cold flow" configuration. *NASA*

Since 1957, Nellis AFB has been famous as the home to the Thunderbirds, the US Air Force aerobatic demonstration team, and since 1975, Nellis and the weapons school have hosted Red Flag, a series of advanced aerial combat exercises. Held several times annually, Red Flag exercises consist mainly of US Air Force units and aircraft but also include assets from other American services and from the air forces of allied nations. During Red Flag, pilots fly realistic combat scenarios in which they are challenged by—and required to undertake simulated combat with—aircraft from US Air Force "aggressor" squadrons that simulate the tactics of potential enemy air forces (see chapter 10).

The Experimental Engine Cold Flow (XECF) experimental nuclear rocket engine assembly was installed at the Nuclear Rocket Development Station in Jackass Flats in 1967. The XECF was similar to the NERVA engine system tested in 1966 but came closer to an operational flight configuration. *NASA*

The size of the huge NTTS range provides a realism to Red Flag operations that would be impossible in a more confined military operating area. Pilots are encouraged to utilize the entire 4,680 square miles for planning their simulated operations—so long as they don't violate the forbidden air space over that mysterious place that they call Dreamland, and which the outside world knows as Area 51.

CHAPTER 2
THE SKUNK WORKS GOES TO PARADISE

ON A BLUSTERY DAY in April 1955, Clarence "Kelly" Johnson of the Lockheed Corporation stepped out of an airplane and felt the crunch of Nevada desert gravel beneath his feet. The cold of winter had long passed, and the oppressive, oven-like heat of summer had not yet enveloped the place.

He looked around at the barren peaks of the Pahranagat Range to the east, the Groom Range to the north, Pahute Mesa to the west, and an endless stark nothingness to the south. Aside from the men traveling with him, nothing moved. There was no sound but that of the breeze in the Joshua trees. Johnson was in the middle of nowhere, and nothing could have pleased him more.

It was, he knew, the perfect place to test an airplane that was so secret that nearly no one at Lockheed, other than those directly involved, would even know about it until well after it started to fly.

Groom Lake was one of a number of sites that Johnson and his team had investigated as a possible testing location for their mystery ship. In November 1954, Johnson had sent Lockheed test pilot Tony LeVier and manufacturing foreman Dorsey Kammerer out to scour the entire desert Southwest for a remote location where an undisclosed "black" airplane could be tested without being seen.

Even the site search was shrouded in secrecy. LeVier and Kammerer took off in a Beechcraft Bonanza, telling everyone that they were going on a hunting trip. They were misleading only in their explanation of the quarry that they were hunting.

As Johnson later wrote in his memoirs, the Groom Lake site had come as a suggestion by US Air Force Colonel Osmond J. "Ozzie" Ritland, who had once worked there during early nuclear testing. Beginning in 1950, Ritland had organized and commanded the 4925th Test Group (Atomic), which had been the unit responsible for the air-dropped nuclear weapons tested at the NTS. Ritland knew the Nevada desert—and of a place then called Watertown landing strip at Groom Lake—well.

Lockheed's "customer" for this still-notional airplane had already promised to make arrangements to acquire this place as a test site, pending Johnson's personal inspection and his thumbs-up on its suitability. The customer was the CIA, in the person of the man

who would be the CIA program manager for this project, Richard Mervin Bissell, Jr., a Yale-educated economist and protégé of high-profile diplomatic troubleshooter and one-time ambassador to the Soviet Union, Averell Harriman. Bissell had run covert special operations for the Office of Strategic Services (OSS) during World War II and had recently been recruited into the CIA, where he helped engineer a coup in Guatemala early in 1954. He was now serving as special assistant for planning and coordination to Director of Central Intelligence (DCI) Allan Dulles.

"One of our first tasks was to find a base from which to operate," writes Johnson in his memoirs. "The Air Force and CIA did not want the airplane flown from Edwards AFB or our Palmdale plant in the Mojave Desert. So we surveyed a lot of territory. There are many dry lakes in and around Nevada, and the lakebeds are generally quite hard, even under water in the rainy season. A site near the nuclear proving grounds seemed ideal, and Bissell was able to secure a presidential action adding the area to the Atomic Energy Commission's territory to insure [sic] complete security."

"Kammerer and I flew to what would be the test base," Johnson recalls. "I had an Air Force compass, and he had some surveying equipment for use on the ground. Kicking away some of the empty .50-caliber shell cases and other remnants of target practice, we laid out the direction of our first runway. A road had to be laid out, hangars constructed, office and living accommodations built, and other facilities provided."

This is how the middle of nowhere became the nexus of black airplane development in the United States. Just as the place later became iconic, the man who started it on that road was already an icon.

Clarence Leonard "Kelly" Johnson is one of the true legends of world aviation technology. In any short list of the greatest airplane designers of all time, Kelly Johnson's name will be found at or near the top. His straight-forward, no-nonsense approach to aircraft design produced not only successful commercial aircraft, but also several planes that were among the most technologically advanced military aircraft in the world at the time he and his team created them.

Born in 1910, Johnson grew up around Ishpeming, Michigan, and graduated from the University of Michigan in 1932 with a degree in aeronautical engineering. The following year, he began his career as a designer at the Lockheed Aircraft Company, working under the legendary chief engineer Hall Hibbard. He helped to design the Lockheed Orion single-engine transport and the Electra twin-engine airliner, which first flew in February 1934 and evolved into a long series of transport, patrol, and attack aircraft. Kelly quickly acquired a reputation as an ingenious engineer, and in 1937, he earned the Sperry Award for his design of the Lockheed-Fowler flap used in the control surfaces of aircraft.

In the late 1930s, Johnson directed the engineering effort that led to the design of the revolutionary Lockheed P-38 Lightning, which was to be the top American fighter in the first part of World War II. All of the top United States Army Air Force (USAAF) aces flew the Lightning.

In 1943, when Lockheed was assigned the task of building what was to be America's first operational jet fighter, Kelly Johnson and a hand-picked team were virtually sealed into a canvas-roofed building next to the Lockheed wind tunnel in Burbank, California, and told to design and build a prototype in 180 days. This was the beginning of the Lockheed Advanced Development Projects (ADP) office, which continues to be known, because of the conditions in its original home, as the "Skunk Works." The name was taken from the place where Al Capp's cartoon character Li'l Abner distilled his Kickapoo Joy Juice.

Kelly Johnson's Skunk Works had the prototype XP-80 jet fighter ready for engine tests in 139 days, and two years later it was the standard jet fighter in the USAAF. One of Johnson's means of facilitating a smooth work flow and instant communications at the Skunk Works was to have the engineers located no more than fifty feet from the place where prototypes were assembled.

After World War II, Johnson and his Skunk Works team designed a series of high performance aircraft, including the remarkable Mach 2 F-104 Starfighter—known as the "missile with a man in it"—which would earn Johnson the Collier Trophy in 1958.

"The Angel from Paradise Ranch," Lockheed's Article 341, the first U-2, is seen here at Groom Lake in July 1955 around the time of its first flight. *Lockheed*

This photo was taken inside what later became Area 51, circa 1959, showing Watertown landing strip and the dry bed of Groom Lake at the top. *USGS*

Late in 1954, half a year after the February 1954 first flight of the first XF-104, Kelly Johnson was summoned to Washington, DC.

These were the chilly early years of the Cold War. The Soviet Union had exploded its first hydrogen bomb, a 400-kiloton weapon, on August 12, 1953, and had followed up with a 28-kiloton, boosted-fission plutonium atomic bomb just eleven days later. A year later, the Soviets conducted eight atmospheric nuclear tests during October 1954 alone. Meanwhile they were building long-range aircraft and missiles to deliver these weapons against the United States. There was a pervasive fear among American leadership, all the way up to President Dwight Eisenhower, that a "nuclear Pearl Harbor" was a very real possibility. Indeed, a RAND Corporation study had suggested that a surprise attack by the Soviet Union could take out 85 percent of the Strategic Air Command (SAC) bomber force.

In July 1954, Eisenhower asked his science advisor, James Killian of MIT, to organize technological capabilities panels to provide advice on how to obtain intelligence data about Soviet activities. The Air Force and the CIA could monitor the fallout from the nuclear tests, squint at fuzzy photographs of the Myasishchev M-4 "Bison" intercontinental bomber, and fly reconnaissance missions along the perimeter of the Soviet Union, but they were blind to the details that could come only from surveillance operations *over* the Soviet Union.

The CIA began working on what they would call the "overhead reconnaissance project." As steps were being taken to develop a reconnaissance aircraft with sufficient range to overfly the huge land mass of the Soviet Union—and a sufficient service ceiling

to avoid being shot down—the first step in the overhead reconnaissance project turned the clock back to an eighteenth century technology: balloons. After a 1951 top secret RAND study of using weather balloons equipped with cameras for extremely high altitude, long-range reconnaissance missions, the Air Force initiated Project Genetrix. Some of the remotely controlled camera technology developed for Genetrix would prove useful a few years later in the Corona spy satellite program.

The quest for such a high-altitude reconnaissance aircraft began in July 1953, as the Wright Air Development Center of the Air Research and Development Command (ARDC) at Wright-Patterson initiated a feasibility study called Project Bald Eagle. This ambitious research was aimed at a notional aircraft called Weapon System MX-2147, which would have a range of 1,750 miles and a service ceiling of at least 60,000 feet (later extended to 70,000 feet), substantially higher than the ceiling of then-existing Soviet interceptors.

Bald Eagle contracts were issued to Bell Aircraft, Fairchild Corporation, and the Glenn L. Martin Company, which had been previously studying such aircraft. Bell's Model 67 aircraft was ordered under the designation X-16, suggesting, for reasons of obfuscation, that it was merely a research aircraft. Fairchild's M-195 was apparently developed no further than an initial concept. Martin proposed its Model 294 aircraft based on their B-57B twin-engine medium bomber, which would be designated as RB-57D. The idea was to extend the wingspan from 64 to 109 feet, which would help it to meet the MX-2147 specifications. The highest promised service ceiling of the three proposed aircraft was the X-16's 69,500 feet.

Part of the reason that the US Air Force favored the Martin bomber was that it was adapted from the English Electric Canberra bomber. In 1953, a specially modified reconnaissance-variant Canberra had secretly overflown the Soviet intercontinental ballistic missile (ICBM) test range at Kapustin Yar. Though the Canberra had been attacked by Soviet fighters, it had completed the mission.

Lockheed was not issued a request for a MX-2147 proposal, but John H. "Jack" Carter, a retired US Air Force officer now at Lockheed, learned about Bald Eagle and suggested to management that the company should submit an unsolicited proposal. Lockheed President Robert Gross and Chief Engineer Hall Hibbard agreed and turned the job over to Kelly Johnson and his Skunk Works team. (In fact, their Starfighter had also originated as an unsolicited proposal.)

As Johnson wrote in his memoirs: "The airplane would have to fly at an altitude above 70,000 feet so vapor trails would not give away its presence, have a range better than 4,000 miles, have exceptionally fine flight characteristics, and provide a steady platform for photography with great accuracy from this high altitude. It would have to be able to carry the best and latest cameras as well as all kinds of electronic gear for its own navigation, communication, and safety."

The aircraft was the Lockheed Model CL-282, which had a fuselage that was adapted from the Starfighter, but with very long, high-aspect-ratio wings that were like a glider's. Because the aircraft could glide during part of its flight profile, fuel could be saved and the range could be greatly extended.

Johnson also designed the aircraft with other features borrowed from sailplanes: landing skids on the fuselage centerline; a detachable, wheeled carriage for take-off; and outrigger wheels at the wingtips. The prototype and production aircraft would, however, be equipped with retractable centerline landing gear rather than skids.

In April 1954, Lockheed formally proposed the CL-282 to the US Air Force. Charles "Bud" Wienberg, who was on the development staff at Air Force headquarters, attended the presentation. He later told Donald Welzenbach, who coauthored the official CIA history of the project, that General Curtis LeMay, the outspoken commander of the SAC, stood up and stormed out of the meeting, stating that he had "no interest in a plane with no wheels and no guns" and that the "whole business was a waste of time."

On June 7, Johnson got word that the Air Force and the CIA turned the CL-282 down because it was, in Johnson's words, "too optimistic."

However, as the radioactive fallout from the October 1954 Soviet nuclear tests drifted out of Siberia, Johnson got the call from Washington. They were ready for another look at the "too optimistic" CL-282. The call had come from Assistant Secretary of the Air Force for Research and Development Trevor Gardner, a leading figure in US Air Force ballistic missile development, whom Johnson described as "a brilliant engineer in his own right."

In the CIA's official history of the program, authors Gregory Pedlow and Donald Welzenbach recall that the civilians at both the air force and the CIA were more inclined to favor the Lockheed proposal than were the uniformed officers, such as LeMay, and they add that the US Air Force did indeed proceed with acquisition of the RB-57D.

However, other air force officers became early advocates of his proposal, mainly because of Johnson's reputation. These included General Donald Putt, a prewar engineering test pilot who now commanded the ARDC, as well as General Charles Pearre Cabell. A previous chief of Air Force Intelligence, Cabell had served as staff director for the Joint Chiefs of Staff from 1951 to 1953 and was now, though still in uniform, serving as deputy director of the CIA.

On the civilian side, Kelly Johnson had perhaps no greater advocate than Trevor Gardner, but he was a demanding advocate. He put Johnson through hoops to prove his point so that there could be no doubting the merits of his CL-282.

The wings for U-2s are shown here being fabricated at a Lockheed factory in Burbank, California, in the 1950s. *Lockheed*

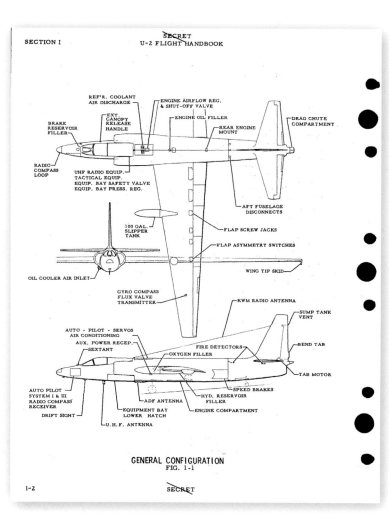

REF'R. COOLANT
AIR DISCHARGE
ENGINE AIRFLOW REG.
& SHUT-OFF VALVE
ENGINE OIL FILLER
DRAG CHUTE
COMPARTMENT
BRAKE
RESERVOIR
FILLER
EXT.
CANOPY
RELEASE
HANDLE
REAR ENGINE
MOUNT
RADIO
COMPASS
LOOP
UHF RADIO EQUIP.
TACTICAL EQUIP.
EQUIP. BAY SAFETY VALVE
EQUIP. BAY PRESS. REG.
AFT FUSELAGE
DISCONNECTS
100 GAL.
SLIPPER
TANK
FLAP SCREW JACKS
FLAP ASYMMETRY SWITCHES
WING TIP SKID
OIL COOLER AIR INLET
GYRO COMPASS
FLUX VALVE
TRANSMITTER
KWM RADIO ANTENNA
SUMP TANK
VENT
AUTO - PILOT - SERVOS
AIR CONDITIONING
AUX. POWER RECEP.
SEXTANT
FIRE DETECTORS
OXYGEN FILLER
BEND TAB
TAB MOTOR
AUTO PILOT
SYSTEM I & III
RADIO COMPASS
RECEIVER
DRIFT SIGHT
ADF ANTENNA
EQUIPMENT BAY
LOWER HATCH
U.H.F. ANTENNA
SPEED BRAKES
HYD. RESERVOIR
FILLER
ENGINE COMPARTMENT

GENERAL CONFIGURATION
FIG. 1-1

A detailed three-view schematic of the U-2 from the CIA's 1959 *Utility Flight Handbook.* CIA

The cover of the CIA's 1959 *Utility Flight Handbook*, was created for use by U-2 pilots. *CIA*

"[Gardner] had assembled a committee of scientists and engineers, and for three days they put me through a grilling as I had not had since college exams," Johnson recalls of his second go-round with the US Air Force and CIA technical staffs. "They covered every phase of the aircraft design and performance—stability, control, power plants, fuels—everything."

In turn, Johnson was invited to lunch at the Pentagon with Air Force Secretary Harold Talbott, CIA Director Dulles, and others. They were taken aback when Johnson promised that Lockheed could "build 20 airplanes with spares for roughly $22 million and have the first one flying within eight months." Don Putt, who was in the meeting, silenced their guffaws of incredulity by reminding them of the ongoing XF-104 program and the fact that the Skunk Works had developed the XP-80 in 139 days.

Another of the early champions of Johnson's CL-282 was the brilliant optical engineer Edwin Land. The inventor of the Polaroid camera and practical polarizing filters for photography, Land chaired one of Killian's technological capabilities panels and would later develop the optics for high-altitude reconnaissance aircraft and spy satellites.

In a November 5, 1954, memo to CIA Director Dulles entitled "A Unique Opportunity for Comprehensive Intelligence," Land called Johnson's proposal not just the best, but the *only* option.

"We believe that these planes can go where we need to have them go efficiently and safely, and that no amount of fragmentary and indirect intelligence can be pieced together to be equivalent to such positive information as can thus be provided," Land wrote. "The Lockheed super glider will fly at 70,000 feet, well out of the reach of present Russian interceptors and high enough to have a good chance of avoiding detection. The plane itself is so light (15,000 pounds), so obviously unarmed and devoid of military usefulness, that it would minimize affront to the Russians even if, through some remote mischance, it were detected and identified."

It also didn't hurt Lockheed that Johnson promised delivery in eight months, and the Bell X-16 prototype was not scheduled for completion before the spring of 1956.

The president himself was convinced. On November 26, the day after Thanksgiving, Dulles told Bissell that Eisenhower had personally authorized the go-ahead for acquiring the Lockheed aircraft to meet the requirements of the CIA's overhead reconnaissance project, which was given the code name Aquatone.

As Bissell recalled in his memoirs, "I was summoned one afternoon into Allen [Dulles]'s office; and I was told with absolutely no prior warning or knowledge that one day previously President Eisenhower approved a project involving the development of an extremely high-altitude aircraft to be used for surveillance and intelligence collection over 'denied areas' in Europe, Russia, and elsewhere. I was to go over to the Pentagon, present myself in Trevor Gardner's office and there with Gardner, General Donald Putt of the Air Force, General Clarence Irvine and others, we were to decide how the project was to be organized and run. The first time I heard Kelly [Johnson]'s name mentioned was in a call put through by Trevor Gardner to Kelly in which he gave him a go-ahead to develop and produce 20 aircraft. We had an almost impossible schedule to meet."

"The secrecy of this project was impressed on me by Gardner," Johnson writes. "I understood that I was essentially being drafted for the job—becoming a 'spook'—the intelligence community's label for their agents. I returned to Burbank with instructions to talk only with Robert Gross and Hall Hibbard. Despite the fact that they had sent me to Washington with instructions not to commit to any new projects because the plant already had several military programs in engineering, they agreed that we must cooperate with this important work."

As Johnson later said, "Security was so strict that after we had submitted our first vouchers for progress payment on the contract [in February 1955], two checks for a total of $1,256,000 arrived in the mailbox at my Encino home. It seemed prudent to establish a special bank account after that."

Gardner himself flew out to Burbank on December 9, 1954, to visit the Skunk Works, to brief Gross and Hibbard, and to formally confirm the contract. Johnson went to work immediately, organizing the project—which he referred to internally as Project X—with

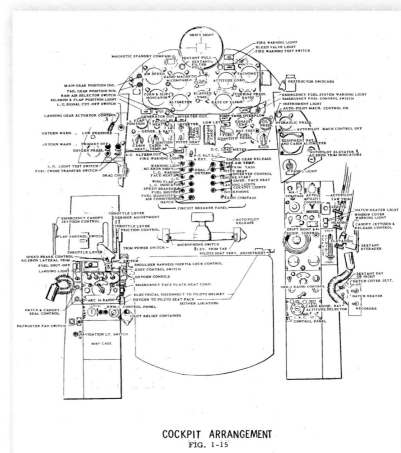

COCKPIT ARRANGEMENT
FIG. 1-15

SECRET

1-29

A detailed look at the U-2 cockpit control panel from the CIA's 1959 *Utility Flight Handbook*. CIA

This CIA U-2, seen parked at Groom Lake circa the late 1950s, carries fictitious NACA tail markings, part of the CIA cover story obscuring the U-2 program. *Tony Landis collection*

twenty-five engineers including himself in the experimental department. Art Viereck was in charge of the shop where the first prototype of the aircraft was essentially handmade.

As this prototype was taking shape in Burbank, the secret flight test location at Groom Lake was also being constructed by a dummy front company that was known only by the name CLJ, these letters being Kelly Johnson's initials. One construction company bidding on the job tried to look up "CLJ" in Dun & Bradstreet and discovered that this "company" didn't even have a credit rating.

"In recruiting mechanics and technical people to work on the project, we named it the Angel from Paradise Ranch," Johnson writes in his memoirs. "Angel because it was such a high-flying airplane, of course, and Paradise Ranch because we thought that would attract people. It was kind of a dirty trick since Paradise Ranch was a dry lake where quarter-inch rocks blew around every afternoon. Actually, in the Skunk Works, we've never had trouble getting workmen to go wherever we've needed them because they know wherever it is the work will be exciting and challenging. And where it's really rough, they are paid a bonus of perhaps 15 percent on top of a good basic salary, plus living expenses. They cannot, of course, take their families with them for security reasons, but they can return home at least once a year when on long assignments."

The aircraft, which Lockheed called the Angel and which the CIA called Aquatone, was assigned the innocuous U for Utility designation: U-2. The U-1 was the De Havilland Otter utility transport, and most U-designated aircraft were light, general, aviation-type aircraft.

The US Air Force designation was just a cover, because at first, the CIA was the actual customer and intended operator. In the CIA's official program history, Pedlow and Welzenbach observe that there was initially "no clear delineation of responsibilities between the CIA and the Air Force," but as the details were worked out, the air force supplied technical support and the Pratt & Whitney J57-P-37 turbojet engines by diverting them away from other programs, such as the KC-135. Meanwhile the CIA provided the funding for the airframe and the reconnaissance cameras out of their secret Contingency Reserve Fund.

As the CIA used the code name Aquatone, the air force referred to the program as Project Oilskin, and each had its own program manager. Bissell held that job for the CIA, while at the air force, Ritland was the first, but he was later succeeded by Colonel Leo Geary.

By July 1955, both Paradise Ranch and the Angel herself were ready. CIA, US Air Force, and Skunk Works personnel were in residence at the ranch, and the completed Angel was disassembled at Burbank and secretly transported 250 miles to the middle of nowhere. In fact, in yet another feature that the U–2 had in common with gliders, it was designed with detachable wings, which made it easy to transport.

The design and appearance of the U–2 prototype had evolved considerably from the earlier Model CL-282 drawings. The aircraft no longer looked like a Starfighter from the side, although if you squinted you could see some similarity. It was the Angel's wings that truly set her apart. Optimized for Mach 2 performance, the Starfighter had a wingspan, frequently called "stubby," of just 21 feet 9 inches. Meanwhile, the U–2's high-aspect-ratio, glider-type wings spanned 80 feet and had an area of 600 feet, nearly three times that of the Starfighter.

As Johnson had promised, less than eight months after the contract was confirmed, the Angel was at on the runway at Paradise Ranch on July 27, 1955, with test pilot Tony LeVier running it up and down in taxi tests as Kelly Johnson looked on.

Like a glider, the wings were so aerodynamic that they actually resisted keeping the aircraft on the ground. On the first day of August, LeVier was routinely taxiing the Angel at about 80 mph. When he began to try the ailerons, he suddenly felt himself not on, but above, the runway.

"I became aware of being airborne, which left me with utter amazement, as I had no intentions whatsoever of flying," LeVier later recalled. "I immediately started back toward the ground, but had difficulty determining my height because the lakebed had no markings to judge distance or height."

"Our first flight was unprogrammed," Kelly Johnson recalls. "The airplane was so light that on his second taxi run it just lifted to a height of about 35 feet [officially 36 feet] before Tony realized he was off the ground. And when he tried to land, the darned airplane didn't want to. It could fly at idle power on the engine. He managed to bounce it down, and in the process bent the tail gear a bit. But we soon had it fixed."

The aircraft was so aerodynamic that it would glide for great distances on its own momentum, defying efforts to force it to land. On August 4, LeVier took off again for an unofficial second flight, with Johnson following him up to 8,000 feet in a C-47 chase plane.

In the early days at Groom Lake, Lockheed and CIA personnel were housed in mobile homes. Note the large air conditioners on the roofs. Groom Lake's first paved runway can be seen in the background. *Tony Landis collection*

Above: Several NACA-marked CIA U-2s can be seen in this view of the busy Watertown flightline at Groom Lake. *Tony Landis collection*

Right: Maintenance work is shown being done on a CIA U-2 at one of the Groom Lake hangars. Even today, much of what happens here happens at night. *Tony Landis collection*

Maintenance At Night On U2.

"The airplane flew beautifully, but again Tony had trouble on landing," wrote Johnson. "He came in tail high and the plane porpoised badly. He made five other attempts before I could talk him down. We discovered the airplane makes a fine landing when the tail wheel hits at the same time or slightly ahead of the main gear."

Luck was with them on timing, as within minutes of the successful landing, a cloudburst exploded over Groom Lake. The dry lakes of the Nevada desert rarely get more than a trace of rain, scarcely more than a few inches annually, which is why they are earn the appellation of *dry* lakes. However, late in the afternoon of August 4, the level expanse of Groom Lake was flooded with at least two inches of water.

On August 8, 1955, the Angel made her official debut flight over Groom Lake. On hand at Paradise Ranch were the dual program managers, Ritland and Bissell, as well as Garrison Norton, an assistant to Trevor Gardner, and other CIA and air force brass. As they squinted into the hot summer sky, Tony LeVier took the Angel up to 32,000 feet. Over the

The weather office at Groom Lake was staffed by Air Force and CIA meteorologists. The calendar is turned to April 1957. Note the huge air conditioning vent. *Tony Landis collection*

This photo shows a fire engine, circa 1950s, parked near the Watertown landing strip control tower. *Tony Landis collection*

coming weeks, LeVier explored the Angel's flight characteristics, taking it up to its design top speed of Mach 0.85 and to gradually higher altitudes. He reached 52,000 feet on August 16 and 65,600 three weeks later on September 8.

To support these flights, which were higher than aircraft had yet operated, the U-2 required low-volatility, low-vapor-pressure fuel that would not "boil off" under reduced atmospheric pressure. This problem was solved with the Shell Oil Company's LF-1A (or JP-TS for Thermally Stable), later called JP-7.

As time went on, the first Angel was flying above 70,000 feet, and she was joined by more U-2 aircraft. With a thumbs up from the customers, Lockheed received a contract for a production series of aircraft, of which forthy-eight were assigned the US Air Force tail numbers 56-6675 through 56-6722. Many of the earliest U-2s had carried National Advisory Committee for Aeronautics (NACA) markings and tail numbers at various times.

Above: This photo shows part of the operational complex at Groom Lake, with the dry lakebed in the foreground. The SP-1 radar dish was used for weather reporting. *Tony Landis collection*

Right: Preparing maps for U-2 flights. The gray shapes indicated on the large map of the United States in the background are the restricted air space of Military Operating Areas. *Tony Landis collection*

By now, the air force had come to seriously believe in the U-2. In their official history of the program from the CIA perspective, Pedlow and Welzenbach note that during 1955, the air force lobbied President Eisenhower to take complete control of the U-2 program, but the president insisted that overflights of Soviet territory should not be conducted by the military.

Early in 1956, however, Curtis LeMay's SAC did reach an agreement with the CIA by which they would acquire their own U-2s from Lockheed. Under the code name Dragon Lady—a term that was much later applied to the U-2 aircraft themselves—the aircraft were ordered from Lockheed by the CIA, which then transferred title to the air force. According to Pedlow and Welzenbach, through early 1959, the CIA received twenty U-2s and the US Air Force acquired thirty-one.

Now that the Angels were flying, many technical challenges still had to be overcome before flights into the thin atmosphere at such high altitudes could be made.

Above: Even today, U-2s are landed by two pilots: one in the cockpit and one in a chase car to keep an eye on the attitude of the wingtips. The dust being kicked up by this bird indicates that it is landing on the dry lakebed of Groom Lake, not on the paved Watertown landing strip. *Tony Landis collection*

Left: A U-2 parked at Groom Lake awaiting overseas deployment for missions in foreign skies. In 1957, photographs of "NACA 320" were the first of the U-2 released to the public. *Author's collection*

Left: The longer-winged prototype U-2R, marked only with the civilian registration N803X, prepares to land at the Watertown landing strip south of Groom Lake within Area 51. *Author's collectio*

CHAPTER 3
ANGELS IN THE BLUE BOOK

AS THE U-2s WERE routinely operating above 60,000 feet, pilots were in an environment where their blood was as prone to "boiling off" as their fuel. The answer was to not only pressurize the cockpit, but also undertake the long and complex process of developing a high-altitude pressurized "space suit" for the pilots. Though Groom Lake was not yet iconic in folklore as a place where the United States government hid men from outer space, there were men living here who did travel to the fringes of outer space.

The conspiracy theorists were yet to make the association, but the Angels would soon be the first flying objects from Groom Lake to titillate the true believers in extraterrestrial spaceships.

A generation before Bob Lazar popularized the term Area 51 for Groom Lake's secret storehouse of extraterrestrials, the term "flying saucer" filled the headlines and movie posters of popular culture. The official government chronicle of this era was the US Air Force's Project Blue Book. Within its pages, a sudden explosion of media and popular speculation about visitors from outer space would collide with the reality of the Angels from Groom Lake.

The term "flying saucer" had originated during the summer of 1947 when a private pilot named Kenneth Arnold flying over the Cascades of southwestern Washington State observed a "chain" formation of nine flying objects. When he reached his destination, Arnold described the flight pattern of the nine discs as being like that of "saucers" being "skipped across water." The news media seized upon his choice of words and the term was born.

There were several other such "sightings" during that summer, but the one which resonated most prominently among the later conspiracy theorists of the Bob Lazar era happened on July 4, just a week after Arnold coined the iconic nickname. Something was reported to have crashed into the desert about 75 miles northwest of Roswell, New Mexico. This was covered in the media at the time, though it was not until much later that reports began to circulate about there having been extraterrestrial beings at the crash site.

In the 1950s, the flood of articles, books, and movies about flying saucers rivaled the volume of the similar flood of speculation about Area 51 and Roswell that was to come

At Yucca Flat, across the hills from the future Area 51, a weather balloon is prepared for launch in January 1957 by personnel from the AEC's Sandia National Laboratories. Under its Project Genetrix, the CIA also used weather balloons to carry reconnaissance cameras over the Soviet Union. A sizable number of the UFOs reported to Project Blue Book turned out to be weather balloons. *NNSA*

in the 1990s. People around the United States were reporting mysterious things in the sky on an almost daily basis. Those who doubted an extraterrestrial origin suggested that the mystery objects might be secret American or secret Soviet aircraft. There was a great deal of concern about a serious violation of American airspace by unknown aircraft among a public for whom the memories of the aerial bombardment during World War II were fresh.

Most, and perhaps all, of these things were certainly natural phenomena, weather balloons, or misidentified airplanes, but even the US Air Force didn't know for sure. General Nathan Twining, then the commander of the Air Materiel Command (AMC) at Wright-Patterson AFB, wrote to Chief of Staff General Carl Spaatz, telling him that he was concerned about the "so-called flying discs" and wanted to see them officially investigated.

U-2 pilots at Groom Lake, wearing civilian clothes, receive a weather briefing from a CIA meteorologist. The lines on the map indicate U-2 flight paths across the United States. There would probably be some correlation between these flight paths and 1957 high altitude UFO sightings reported to Project Blue Book. *Tony Landis collection*

In the memo, he told Spaatz and Brigadier General George Schulgen of the Air Force Office of Research & Development that he had reviewed the reports that had come in and that he had discussed the subject with personnel of the AMC Engineering Division and the Air Institute of Technology. It was their opinion that the "phenomenon is something real and not visionary or fictitious."

Twining conceded that there was a "possibility that these objects are of domestic origin—the product of some high security project not known to [the Headquarters Air Staff] or the [AMC]."

The general's suggestion that there might be "some high security project" whose level of security was beyond that for which even he had been cleared raised some eyebrows at the time, but the focus of popular culture for the next generation would be on the otherworldly and not yet on homegrown mystery airplanes.

He went on to advocate "a detailed investigation of this matter . . . in order to more closely define the nature of the phenomena," which was undertaken by his successor, General Joseph McNarney, after Twining departed in October after less than two years in the post. (Between 1953 and 1957, Twining served as chief of staff of the air force, during which time he was well aware of the U-2, and presumably, there were no "projects not known" to him.)

Meanwhile, in December 1947, the Air Staff ordered the Air Technical Intelligence Center (ATIC), an AMC component, to look into the matter. Thus began Project Sign, which examined more than 200 sightings through early 1949. This investigation concluded that "the possibility exists that some of the incidents may represent technical developments

Working in an air control center, US Air Force Major Trumble and an unidentified captain monitor flights in the restricted air space above the Nevada Test Site and the Groom Lake complex. *NNSA*

A natural metal U-2A in US Air Force markings is shown here at moderate altitude. Because U-2s operated secretly at altitudes above which no other aircraft was known to operate, the reflection of sunlight on their silvery wings was occasionally reported as a sighting of a UFO. *Author's collection*

As is shown by these front page headlines from July 8 and 9, 1947, a great deal of media attention surrounded the crash of an alleged "flying saucer" near Roswell, New Mexico. Though front page news at the time, the Roswell incident faded quickly. Strangely, it was virtually ignored by both Project Blue Book *and* by the UFO conspiracy community for decades. *Author's collection*

far in advance of knowledge available to engineers and scientists of this country" and went on to add simply that "no facts are available to personnel at this Command [AMC] that will permit an objective assessment of this possibility."

This inconclusive conclusion only fanned the flames of popular interest in the topic. The media firestorm swirling around the flying saucer sensation had given it a life of its own.

Two of the air force staff officers who were instrumental in seeing that the ATIC investigation of aerial phenomena continued were General Don Putt and General Charles Cabell, both of whom were later important supporters of the U-2 project. In February 1949, at their urging, a new program, called Project Grudge, picked up where Project Sign had left off. The Grudge report, issued in August 1949, concluded that "these flying objects constitute no threat to the security of the United States" and that the things that people claimed to be seeing in the sky were "the result of misinterpretations of conventional

Left: Plotting the Nevada Test Site and Groom Lake air space in March 1955, Colonel Fackler checks the time as Major DeVries locates the positions of aircraft. *NNSA*

Below: Radarscope images from Project Blue Book File 4517 track a UFO observed over MCAS El Toro on the night of November 11-12, 1956. A U-2 could easily have been in that air space, flying higher than Marine Corps aircraft could have climbed. *National Archives*

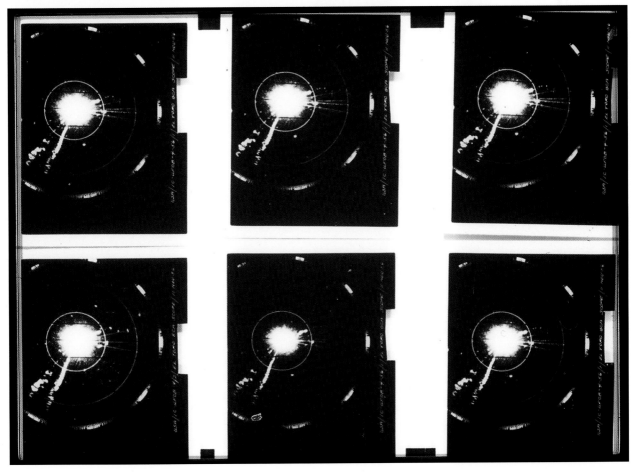

objects, a mild form of mass hysteria or war nerves, and individuals who fabricate such reports to perpetrate a hoax or to seek publicity."

Even as Grudge sought to dismiss the "misinterpretations of conventional objects" and the "mild hysteria," the public imagination had not been satisfied. Within popular culture the notion that these flying saucers were from outer space had gained serious traction as the most vivid part of the narrative. It didn't help that President Harry Truman, hoping to allay concern that there were Soviet aircraft in American air space, famously told a press conference on April 4, 1950, "I can assure you the flying saucers, given that they exist, are not constructed by any power on Earth."

After Project Grudge wound down, ATIC continued to collect reports, and in 1951, Captain Edward Ruppelt was assigned to this task. An air force reservist called to active

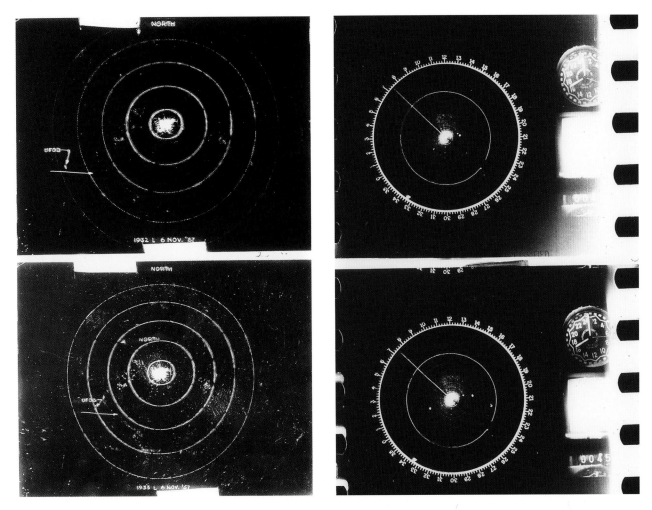

duty because of the Korean War, Ruppelt took his job of organizing and collating reports seriously. He even coined an acronym to describe those things in the sky. He simply called them "unidentified flying objects," or "UFOs." Who could have predicted that this term would become one of the most ubiquitous in popular culture?

In March 1952, on the initiative of General Cabell—now heading Air Force Intelligence—UFO research was transferred from ATIC to a new entity at Wright-Patterson AFB that was known as the Aerial Phenomena Group (APG). Here, it was reorganized under a new program called Project Blue Book. It is unclear whether Cabell really believed that unknown things were flying in American air space or if he found flying saucers to be a useful obfuscation of secret programs of which he was well aware.

Ruppelt headed Blue Book until 1954, but the investigation would continue through 1969 under a series of other directors.

After 1955, a proportion of unidentified sightings that were reported to Project Blue Book were actually U-2s. As Gregory Pedlow and Donald Welzenbach write in the CIA's own official history of the U-2, "High-altitude testing of the U-2 soon led to an unexpected side effect—a tremendous increase in reports of unidentified flying objects (UFOs). In the mid-1950s, most commercial airliners traveled at altitudes between 10,000 and 20,000 feet and military aircraft like the B-47s and B-57s operated at altitudes below

Two images on the left: As noted in Project Blue Book File 5178, this fast-moving object labeled as "UFO B" was captured on radar over Whiteman AFB on November 6, 1957. *National Archives*

Two images on the right: On May 31, 1963, this group of UFOs was observed over the Atlantic Ocean. Could they have been U-2s? *National Archives*

One of the most intriguing images in the Project Blue Book archive is this one from File 9337. The case involved a saucer-shaped craft that a man named John Reeves observed on March 3, 1965, near Brooksville, Florida. When the file was turned over to the National Archives, the image had been censored. It is unknown what lies beneath the clipped paper, but the expression on the face of the investigating airman speaks volumes. *National Archives*

40,000 feet. Consequently, once U–2s started flying at altitudes above 60,000 feet, air traffic controllers began receiving increasing numbers of UFO reports."

Visually the aircraft were hard to identify because they operated in the troposphere, leaving no tell-tale contrail, just an isolated reflection of sunlight. Pedlow and Welzenbach go on to explain that reports of U–2s as UFOs "were most prevalent in the early evening hours from pilots of airliners flying from east to west. When the sun dropped below the horizon of an airliner flying at 20,000 feet, the plane was in darkness. But if a U–2 was airborne in the vicinity of the airliner at the same time, its horizon from an altitude of 60,000 feet was considerably more distant, and being so high in the sky, its silver wings would catch and reflect the rays of the sun and appear to the airliner pilot 40,000 feet below, to be fiery objects. Even during daylight hours, the silver bodies of the high-flying U–2s could catch the sun and cause reflections or glints that could be seen at lower altitudes and even on the ground. At this time, no one believed manned flight was possible above 60,000 feet, so no one expected to see an object so high in the sky."

Had the original black aircraft from Area 51 actually been painted black—as were later U–2s—many of these UFO sightings never would have happened.

Pedlow and Welzenbach add that "Blue Book investigators regularly called on the [Central Intelligence] Agency's Project Staff in Washington to check reported UFO sightings against U–2 flight logs. This enabled the investigators to eliminate the majority of the UFO reports, although [Blue Book] could not reveal to the letter writers the true cause of the UFO sightings."

By the late 1960s, Project Blue Book was under increasing criticism from conspiracy theorists for covering up the true—and possibly extraterrestrial—nature of UFOs and from UFO skeptics for being a big waste of time and money. In 1966, with this public controversy ongoing, the Air Force Scientific Advisory Board (SAB) commissioned the University of Colorado to take a definitive look at the UFO phenomenon. This effort was headed by Dr. Edward Condon, a nuclear physicist who had worked on the Manhattan Project during World War II and who was later the director of the National Bureau of Standards. The contract provided that the planning, direction, and conclusions of the Condon study were to be conducted wholly independently of the US Air Force. To avoid duplication of effort, the air force was ordered to furnish the project with the records of its own earlier work and to provide the support of personnel at air force bases when requested by Colorado field teams.

Condon's 1,400-page report, issued in 1969, concluded that the US Air Force had uncovered no evidence of extraterrestrial visitors. Condon had a number of vocal critics, including the astronomer J. Allen Hynek, who had been a consultant to the earlier air force investigations, but by and large, the scientific community accepted Condon's conclusions. So, too, did the media, and two decades of intense public interest in UFOs withered into hibernation to await new interest by later generations.

The US Air Force, in turn, used Condon's conclusion as an opportunity to terminate Project Blue Book once and for all. As noted in the official US Air Force fact sheet, Project Blue Book concluded that "no UFO reported, investigated and evaluated by the Air Force was ever an indication of threat to our national security; There was no evidence submitted to or discovered by the Air Force that sightings categorized as 'unidentified' represented technological developments or principles beyond the range of modern scientific knowledge; and There was no evidence indicating that sightings categorized as 'unidentified' were extraterrestrial vehicles."

They were careful not to say "*declassified* technological developments or principles."

Officially, Blue Book had investigated a total of 12,618 UFO sightings during its seventeen years, identifying all but 701 as misinterpretations of conventional objects. According to Pedlow and Welzenbach, *most* of the "unidentified" ones were secret aircraft from this world. The other unexplained sightings remain unexplained.

General Charles Pearre Cabell played an important role both in the UFO investigations of the 1950s and in the later operational deployment of the U-2. In 1953, he became deputy director of the CIA. *USAF*

General Don Putt, an experienced test pilot, later served as deputy commander for intelligence at the Air Technical Service Command and as commander of the Air Research and Development Command. In 1954, he took over as director the chief of staff's Scientific Advisory Board. Like Charles Cabell, he played a role in UFO investigations and was an important advocate of the U-2 program in its formative years. *USAF*

CHAPTER 4
ANGELS IN RED SKIES

IN 1956, AS FLYING SAUCERS were appearing regularly on the covers of pulp magazines, the US Air Force and the CIA were preparing to deploy U-2s overseas for operational missions.

Meanwhile, Project Genetrix had come to an end after only two months. Nearly fifty cameras and film capsules had been recovered, but most of the balloons had not successfully drifted across the entire breadth of the Soviet Union. Ironically one of the most useful results from having reconnaissance balloons masquerading as weather balloons was the weather data that was gathered. A great deal was learned from high altitudes winds over the Soviet Union that would be vital in planning the U-2 missions.

Meanwhile the cameras that had been developed for the U-2 were, like the airplane itself, breaking new ground technologically. They had to. Existing aerial cameras had good resolution when taking pictures from 30,000 feet, but the U-2 would be flying twice that far from the subjects of its cameras. The Connecticut-based Perkin-Elmer company, an existing maker of aerial cameras, had developed high-acuity K-38 cameras, but they needed to scale it down to meet the U-2's 450-pound payload limit. The result was the A-1 system that consisted of a pair of K-38s with a K-17 as a back-up camera.

Of course, in the context of Cold War geopolitics, the notion of overflying the Soviet Union involved more dimensions than cameras, altitude, and weather. Even as he approved Project Aquatone at each step, President Eisenhower was well aware of the politics of violating the air space of another superpower. He voiced this opinion directly to CIA Director Allen Dulles and Richard Bissell, as well as through his Defense Liaison Officer Colonel Andrew Jackson Goodpaster.

Back on July 21, 1955, less than a week before the Angel went airborne for the first time, Eisenhower had made his famous Open Skies proposal to the Soviet Union in which he offered to allow the Soviets to overfly American facilities if they would reciprocate. Eisenhower had considered their rejection of this proposal as a rationale for continuing the Aquatone program, but a year later, he hesitated, fearing the reaction that would come if the Soviets detected a U-2, or even worse, if one of the aircraft crashed or was shot down.

The CIA, especially Dulles, insisted that this was unlikely and that it was worth the risk when balanced against the vital intelligence that was likely to be gained by the overflights.

Goodpaster, who served as Eisenhower's point of contact with the CIA with regard to overhead reconnaissance and who sat in on the president's meetings on the subject, recalled Dulles's almost nonchalant attitude on the subject. In the early 1980s, he told Smithsonian historian Michael Bechloss that "Allen's approach was that we were unlikely to lose one. If we did lose one, the pilot would not survive....We were told—and it was part of our understanding of the situation—that it was almost certain that the plane would disintegrate and that we could take it as a certainty that no pilot would survive and that although they would know where the plane came from, it would be difficult to prove it in any convincing way."

Just as Genetrix operated under the weather balloon cover story, the CIA U-2s operated under the cover story of being weather reconnaissance aircraft ordered by National Advisory Committee for Aeronautics, which explains why the early ones bore NACA markings. These aircraft were assigned to the US Air Force 1st Weather Squadron, Provisional, with the latter qualification added because "provisional" squadrons could exist outside a routine command structure.

The pilots, meanwhile, were fighter pilots who held reserve, rather than regular, commissions. They resigned these commissions in order to be hired as civilians by the CIA and then masqueraded as US Air Force pilots flying with the newly created provisional squadron.

Detachment A of the 1st Weather Squadron began deploying overseas to England in April 1956 but was redeployed to the US Air Forces in Europe (USAFE) base at Wiesbaden

Just as precious to the Soviets as Area 51 was to the Americans, and for a time just as secret, was the ICBM test facility at Tyuratam. As Area 51 is in the arid Nevada desert, this site is located on the desert steppes of Kazakhstan. Another thing that both had in common in 1957 was U-2s overhead. This photo, showing an ICBM on the pad, was taken by a U-2 in August. Later in 1957, Tyuratam garnered international notoriety as the Soviets launched the first earth-orbiting satellite, Sputnik I. *CIA*

A natural metal U-2A flies at 65,000 feet, which seems like the edge of space. This U-2A, later reconfigured as a U-2B, had a long career with the CIA before being transferred to the Air Force Systems Command. Upgraded to U-2C standard for the 100th Strategic Reconnaissance Wing in 1968, it was painted black in 1970 and saw service over North Vietnam. *Lockheed*

Allen Dulles was a lawyer and banker who joined the OSS during World War II to serve as a clandestine operator and as Switzerland station chief. Between 1953 and 1961, he was the director of the CIA. He advocated the CIA having its own aerial reconnaissance capability.
National Archives

in West Germany in June. According to Pedlow and Welzenbach, this move was to "avoid arousing further [public] reaction [to the U-2's arrival] in the United Kingdom." They were quickly moved to another location near the East German border whose name is redacted in the copy of the Pedlow–Welzenbach document available to the author. Here the aircraft were reengined with more powerful J57-P-31 engines and were redesignated as U-2Cs, a development discussed in more detail in the following chapter.

While waiting for Eisenhower's final go-ahead for flights over the Soviet Union, the first U-2 missions, over East Germany and Poland, were flown on June 20. At the CIA, Richard Bissell and General Charles Cabell, now the agency's deputy director of central intelligence (DDCI), were eager to begin flights over Soviet territory, but Eisenhower insisted on a face-to-face briefing for German Chancellor Konrad Adenauer. The two men flew to Bonn personally.

On July 4, less than one year after the U-2's debut flight, the aircraft was flying an operational mission over the naval shipyards at Leningrad on its first Soviet Union mission. The following day, a U-2 overflew Moscow itself for the first and only time, then flew 125 miles farther east, looking down at the facility at the Zhukovsky Airfield at Ramenskoye where the Myasishchev M-4 Bison bombers were tested.

In a later conversation with Donald Welzenbach, Richard Bissell recalled briefing Dulles about Leningrad and Moscow having been overflown in the first twenty-four hours of the surveillance program. If Dulles was nonchalant, Bissell was almost cavalier.

"Oh my Lord," Dulles said. "Do you think it was wise the *first* time?"

"Allen," replied Bissell. "The first time is the safest."

In turn, Dulles and Bissell met with Goodpaster to discuss the president's concerns about whether the first overflights had been tracked on Soviet radar. In his July 5, 1956, memorandum for the record, Goodpaster noted that the CIA was authorized to continue the overflights "at the maximum rates until the first evidence of [radar] tracking was received." In a July 1987 interview, Goodpaster told Donald Welzenbach that Eisenhower was prepared to immediately halt the overflights if the U-2s were detected.

As the mission folders in the files of the CIA Office of Special Activities (OSA) show, the Soviets did detect the U-2 but were unable to track it consistently. Indeed, their radar coverage was so spotty that they did not know the aircraft had been over Moscow or Leningrad.

Elsewhere, MiG-15s and MiG-17s were captured by the K-38 cameras as they attempted to reach the Angel. However, the fighters were unable to reach the U-2 at its altitude. Just as the gleaming bare metal belly had made American airliner pilots mistake it for a flying saucer, the same glare made the Angel easy for Soviet pilots to see, even if they could not touch it.

Naturally, the Soviets protested about the overflights—privately, of course, because to admit being overflown would have been embarrassing.

Eisenhower then told Dulles to halt the flights—after eight missions behind the Iron Curtain, including five over the Soviet Union itself—and to tell no one about the U-2 missions who did not already know. The president met with Dulles on July 19, where he told him, according to Goodpaster, that he had "lost enthusiasm" for U-2 overflights of Soviet

President Dwight Eisenhower learned the importance of aerial reconnaissance as the Supreme Allied Commander in Europe during World War II. As president, he authorized the development of the U-2, but he grew increasingly nervous about direct overflights of the Soviet Union. *Eisenhower Library*

Andrew Jackson Goodpaster was a 1935 West Point graduate who served on General George Marshall's staff during World War II. From 1954 to 1961, he was Eisenhower's staff secretary and defense liaison officer. *NATO*

territory, although he did agree for them to continue over Eastern Europe. In an October 3 conversation with Goodpaster, a nervous Eisenhower grumbled that the U-2 operations were "provocative and unjustified."

Despite all of the apparent failures, the U-2 had already achieved an unexpected intelligence coup. Analysts were able to ascertain that not nearly as many Bison bombers were being rolled out as previously feared. Indeed, there was no "bomber gap."

In the fall of 1956, even as the overflights of the Soviet Union were suspended, the attention of the Eisenhower administration and the CIA shifted to the Middle East. Egypt had seized the Suez Canal, then owned and operated by Britain and France. The crisis devolved into open warfare as Israel attacked Egypt in the Sinai Peninsula, and Britain and France attempted to seize the canal by force. The United States remained on the sidelines with Eisenhower demanding a halt to hostilities.

Meanwhile, as Detachment A in West Germany was closed down, the CIA had made arrangements with Turkey, a fellow North Atlantic Treaty Organization (NATO) member, to base U-2 Detachment B there for possible future overflights of the Soviet Union. When the Suez Crisis began, the U-2s were able to supply Eisenhower with timely information about unfolding events. Later in 1956, after the Soviets intervened in Hungary to crush a rebellion against their dominance of Eastern Europe, Eisenhower agreed to a resumption of limited overhead reconnaissance by Detachment B of Eastern Europe, but not of the western Soviet Union.

Some CIA U-2s carried US Air Force markings. Pilots operating in dangerous foreign skies lobbied for the darker paint to make the aircraft harder to be seen by interceptors. *Lockheed*

This aircraft is one of several U-2As that were upgraded to U-2D standard, capable of carrying additional reconnaissance gear or a second crewmember. On its tail, it carries the insignia of the Air Force Flight Test Center at Edwards AFB. *Author's collection*

This image provides a good view of a rotating optical sensor ball that was mounted ahead of a communications antenna fairing of U-2D number 56-6721. *Lockheed*

As it was now obvious that Soviet radar could track the U-2s, the CIA initiated Project Rainbow, an effort to develop means of reducing the U-2's radar cross section (RCS). These developments were early steps toward the basket of technologies known as "stealth," which emerged into prominence a quarter century later.

One of the firms that emerged as a major player in this process was Edgerton, Germeshausen, and Grier (EG&G), a technical consulting firm founded in 1931 by MIT professor Harold Edgerton, a pioneer of high-speed photography. During World War II, EG&G had the imaging technology for Manhattan Project implosion tests, and in the 1950s, they were one of the key support organizations for the nuclear testing program at the NTS.

Over the coming decades, EG&G gradually expanded the scope of their work from engineering to facilities management at secure government locations, especially within the Nellis Range. As Area 51 mythology unfolded late in the century, EG&G was often singled out in various "black airplane" conspiracy theories. For Project Rainbow, their role was that of monitoring the proto-stealth experiments developed by others.

Above: Using his drogue chute to slow him down, a U-2 pilot makes a landing at the Watertown landing strip. Keeping the wings perfectly balanced on landing is an essential acquired art for U-2 pilots. *Tony Landis collection*

Right: A rare photo of a U-2 configured as a bomber. This wind-tunnel model is fitted with a variety of bombs and missiles on underwing pylons. *Author's collection*

The early radar deception experiments conducted at Groom Lake involved radar-absorbing beads on wires strung around the periphery of the U-2 or gluing radar-absorbing "wallpaper" panels to its fuselage. Aircraft this encumbered were called "dirty birds," an appellation readily accepted by Kelly Johnson, who did not like these additions because they interfered with the aerodynamics of his airplane.

This came to a head on April 2, 1957, when a dirty bird, coincidentally Article 341, the U-2 prototype, piloted by top Lockheed test pilot Robert Seiker (sometimes seen spelled as Sieker) crashed in a remote part of the Nellis AFB Range. It was not found for several days. The deadly mishap was traced to overheating caused by wallpaper, which resulted in a stall.

Ultimately, and sadly, given that they had claimed a life, the dirty bird modifications were proven ineffective.

Later RCS testing would be conducted with aircraft held aloft by a crane, or positioned atop a pole or pylon.

Soviet radar notwithstanding, overflights of parts of the Soviet Union east of the heavily populated areas had resumed and were proving extremely useful. During August 1957, under a series of missions conducted under the code name Soft Touch, U-2s brought back significant images of the ICBM test—and future space launch—facilities at Semipalatinsk and Tyuratam in Soviet Kazakhstan.

Russian Colonel Alexander Orlov told a CIA/CSI public symposium in September 1998 that "between March and October [of 1957], Soviet air defense radar picked up five U-2 overflights . . . at altitudes of 19 to 21 kilometers [about 62,000 to 69,000 feet], they were beyond the reach of the Soviet Air Defense Forces' fighter planes and antiaircraft artillery."

A short time later, the U-2s of Detachment C, flying out of Eilsen AFB in Alaska, photographed the nuclear weapons and missile facilities at Klyuchi on Kamchatka Island in the Soviet Far East.

In the span of a few days, Charles Cabell and Richard Bissell were able to put photos on President Eisenhower's desk that showed him the Soviet equivalents of the NTS and Cape Canaveral. If all three men were as amazed to see the launch site at Tyuratam, they were astonished two months later on October 4 when the Soviets launched Sputnik 1 from here.

Also at Tyuratam, the Soviets built the Baikonur Cosmodrome, their manned space launch center, and in April 1961, Yuri Gagarin went aloft to become the first human to orbit the Earth in outer space. Since the turn of the twenty-first century, numerous American astronauts have traveled into space from Baikonur.

Despite the success of Soft Touch, Eisenhower's reticence in the face of Soviet protests—and attempts to shoot down U-2s in international air space of the Black Sea—led to a winding down of deep-penetration overhead reconnaissance missions. On March 7, 1958, the president told Goodpaster to tell Cabell and Bissell to halt the U-2 surveillance flights completely, initiating a ban that would last sixteen months.

Goodpaster sent Dulles and Bissell a memo conveying the president's demand that "every cent that has been available for any project involving crossing the Iron Curtain is to be impounded and no further expenditures are to be made."

At Groom Lake, the flag flies at half staff for Lockheed test pilot Robert Seiker, who was killed when his U-2 crashed on April 2, 1957. *Tony Landis collection*

CHAPTER 5

ANGELS AT THE TURNING POINT

IN THE MONTHS AFTER President Eisenhower ordered CIA Director Dulles to pull the plug on overflights of the Soviet Union, a great many things happened that would affect the future of the U-2. On the operational side, the aircraft were again flying useful missions in the Middle East, this time watching over American troops who intervened in the 1958 Lebanon Crisis and monitoring Soviet ships and submarines in the Mediterranean. A secret deal to base the planes at Peshawar in Pakistan had been concluded, although no aircraft had yet been deployed there.

Meanwhile Detachment C, now based in Japan, had conducted overflights of the Peoples' Republic of China, notably during 1958, when there were fears that the "Red" Chinese might attempt to invade Taiwan (the Republic of China). Later in the same year, the U-2s were also used—reverting to their original cover story—to monitor the progress of Typhoon Winnie as it came over Taiwan. The United States later transferred some U-2s to Taiwan's Republic of China Air Force. These were operated by the Republic of China Air Force (ROCAF) 35th "Black Cat" Squadron, mainly over mainland China, between 1960 and 1974. The ROCAF was the only non-US air force to officially operate the U-2.

On the technical side, the CIA and US Air Force fleets of U-2As and two-seat U-2Bs were being upgraded. They were retrofitted with larger intakes, reengined with the new Pratt & Whitney J75-P-13 turbojet engines, and redesignated as U-2Cs. The J75's 17,000 pounds of thrust permitted a more rapid climb into the troposphere and a stated operational altitude of 74,600 feet.

Meanwhile some earlier model U-2s were upgraded to U-2D standard, capable of carrying additional reconnaissance gear or a second crewmember. Later other aircraft were modified to be capable of being aerially refueled—although the endurance limitations of the aircraft did not rest with the fuel load but with pilot fatigue. Ten hours was shown to be the maximum length of time that a pilot could function at optimum performance. Aircraft with J57 engines retrofitted for aerial refueling were reportedly redesignated as U-2Es, while J75-powered aircraft became U-2Fs. The U-2G designation went to three U-2As, which were modified with arrestor hooks and other equipment for use aboard US Navy aircraft carriers.

A late production U-2R (designated as TR-1 from 1981 to 1992) from the 9th Strategic Reconnaissance Wing at Beale AFB flies over San Francisco's Golden Gate Bridge in 1985. *USAF, Ken Hackman*

Because of fears that the Soviets might soon deploy an interceptor that could threaten the U-2, the pilots wanted something done about that gleaming bare metal belly that made them feel like sitting ducks against the dark sky. Kelly Johnson's engineers had resisted painting the U-2 because the weight of the paint would cost them altitude. With a more powerful engine installed, though, Johnson relented and the aircraft were painted a very dark blue-black.

On the geopolitical side, the Cold War arms race had heated up again. Just as there had once been a "bomber gap," now there were cries in the media of a "missile gap," as the Soviet Union was reportedly piling up ICBMs at a rapid rate. With nothing to refute these reports, and with mounting pressure from Congress to do something, Dwight Eisenhower was compelled to rethink his overflight ban.

One flight was made over Tyuratam on July 9, 1959, but further missions were complicated by an apparent thaw in relations between the Soviet Union and the United States. Vice President Richard Nixon visited the Soviet Union later in July, and Soviet Premier Nikita Khrushchev returned the favor with a thirteen-day visit to the United States. Khrushchev was well received during his trip, which included coast-to-coast stops that were widely covered in the media and culminated in a meeting with President Eisenhower at Camp David. In turn, the two men made plans for another summit conference to be held in the spring of 1960.

Behind the scenes, however, the Soviet Union was working overtime to develop U-2 countermeasures. These included a high-altitude variant of the Yakovlev Yak-25 interceptor (NATO code name Flashlight) that was designated Yak-25RV (NATO code name Mandrake), as well as the high-altitude V-750VN variant of the S-75 Dvina surface-to-air missile (SAM), which was known to NATO as the SA-2 Guideline.

Ominously, considering later events, Colonel William Burke of the CIA's Development Projects Division (DPD) wrote to Richard Bissell on March 14, 1960, that ATIC's "present evaluation is that the SAM (Guideline) has a high probability of successful intercept at 70,000 feet providing that [radar] detection is made in sufficient time to alert the [SAM launch] site."

Meanwhile the Soviet counterpart to the CIA, the KGB (Komitet Gosudarstvennoy Bezopasnosti, or Committee for State Security), was also pulling out all the stops to get their hands on information about the U-2. Their agents and stringers haunted the periphery of bases from Turkey to Japan, where U-2s were based. If they had known about the secret world at Groom Lake, they would have had agents climbing the hills of the Pahranagat Range to have a peek, just as later-generation black airplane buffs would be doing.

With regard to intelligence about the U-2, conspiracy theorists often make mention of a former US Marine who had worked as a radar operator at Atsugi in Japan and who

A CIA U-2 pilot identified as Francis Gary Powers poses in his high altitude flight suit with an early U-2B, one of several such aircraft fitted with the ventral antenna fairing atop the fuselage. *Lockheed*

Above: When Francis Gary Powers was shot down near Sverdlovsk on May 1, 1960, the CIA quickly moved to reinforce a cover story that his U-2 was a NASA weather plane that had gone off course. The aircraft was painted with a fictitious tail number, and this photo was released to the media. *NASA*

Left: Kelly Johnson shares a photo-op with Francis Gary Powers after his release from Soviet custody in 1962. Johnson hired Powers to work for Lockheed. *USAF*

defected to the Soviet Union in October 1959. His name, which would enter the annals of infamy in Dallas, Texas, on November 22, 1963, was Lee Harvey Oswald. In fact, the future assassin of John F. Kennedy had been no closer to the U-2 than his radar scope, and the Soviets showed no immediate interest in debriefing him.

By the spring of 1960, the CIA had developed an ambitious plan to utilize their U-2 launch site in Pakistan. They would fly 3,700 miles, cross the breadth of the Soviet Union, photograph ICBM facilities at Plesetsk and Sverdlovsk, and land in Norway. One mission, designated as Square Deal, had taken place on April 10, and another, designated as Grand Slam, was scheduled for May 1.

In retrospect, it seems counterintuitive, especially considering Eisenhower's cautiousness about U-2 flights, that a mission would be scheduled for May Day. It was a major Soviet holiday, and it came just two weeks before the president was scheduled to sit down in Paris with Nikita Khrushchev on May 16 for their summit conference.

On the morning of May 1, Francis Gary "Frank" Powers, a veteran of 27 U-2 missions, climbed into a U-2C, Article 360, tail number 56–6693. He took off from Peshawar and flew north into the Soviet Union. He crossed Kazakhstan, photographed Tyuratam, entered

An October 14, 1962, photograph was taken by a U-2 of a Soviet MRBM launch site at San Cristobol in Cuba. It was used as evidence by which President John F. Kennedy ordered a naval quarantine of Cuba. *DOD*

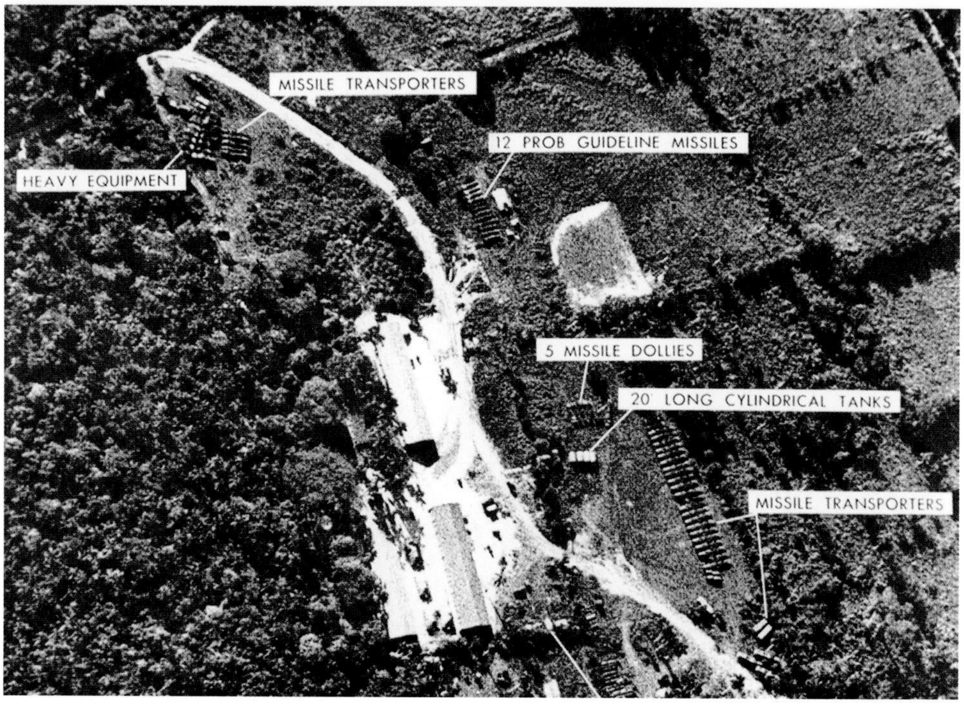

MRBM FIELD LAUNCH SITE
SAN CRISTOBAL NO 1
14 OCTOBER 1962

ERECTOR/LAUNCHER EQUIPMENT

TENT AREAS

EQUIPMENT

ERECTOR/LAUNCHER EQUIPMENT

8 MISSILE TRAILERS

CONSTRUCTIO

This U-2 photo shows a Soviet SAM site in Cuba, which would have become operational with SA-2 Guideline (S-75 Dvina) missiles used to protect MRBM sites from American air strikes. One U-2 was downed by a SAM during the Cuban Missile Crisis. *DOD*

MISSILE TRANSPORTERS

12 PROB GUIDELINE MISSILES

HEAVY EQUIPMENT

5 MISSILE DOLLIES

20' LONG CYLINDRICAL TANKS

MISSILE TRANSPORTERS

On October 23, 1962, citing U-2 imagery, President John F. Kennedy signs off on the quarantine against Cuba. *Library of Congress*

Russia, and was 70,500 feet over the town of Degtyarsk, about 42 miles west of Sverdlovsk, when he was hit by a V-750VN missile. Powers bailed out and was captured by the Soviets, who made no immediate public mention of the incident.

A U-2 headed for North Vietnam in formation with a DC-130 carrying a Lightning Bug reconnaissance drone beneath its wing. *Al Lloyd collection*

When Powers did not arrive in Norway as planned, a preplanned cover story was released to the media through NASA on May 2 that a weather reconnaissance aircraft was missing on a flight over Turkey. On May 5, Khrushchev went public with a widely reported announcement that an American "spyplane" had been shot down.

For two days, he made no mention of the pilot having survived, but when he did, he announced that Powers had admitted to being a spy. Then the largely intact camera system was shown to the media in Moscow. The CIA, the US Air Force, and the Eisenhower administration were humiliated by their own bogus cover story—not to mention the embarrassment to NASA, who provided the cover story.

In the official CIA history of the U-2, Gregory Pedlow and Donald Welzenbach write that "Richard Bissell and the Development Projects Division had become overconfident and were not prepared for the 'worst case' scenario that actually occurred in May 1960."

Pedlow and Welzenbach paint a picture of the CIA having come to believe its own unrealistic assurances about how long the U-2 could remain invulnerable to Soviet air defenses. They write that as early as 1956, the agency assumed the aircraft would have a useful service life of eighteen to twenty-five months, with Richard Bissell believing that it would become vulnerable before the end of 1957.

Above: In 1981, Lockheed resumed production of the U-2R under the TR-1 designation. The aircraft were built at the company's facility in Palmdale, California. *Lockheed*

Right: Retired Skunk Works chief Kelly Johnson at the rollout in Palmdale of the first newly built TR-1 (U-2R) on July 15, 1981. *USAF, Master Sergeant Paul Hayashi*

Now, however, here it was 1960, and the CIA had still never developed a worst case contingency.

In one ray of light, Kelly Johnson successfully cajoled the Soviets into displaying the entire crashed U-2 publicly.

"Hell, no," he told the media when they displayed the wreckage of a Soviet aircraft, possibly a MiG that was accidentally shot down by a SAM trying to hit Powers. "That's no U-2."

It wasn't, and the Soviets promptly dragged out the real wreckage, which was photographed in great detail. The widely published images allowed the Skunk Works team a close look.

The summit conference in Paris in mid-May, which also involved the British and French, turned sour when Eisenhower refused to apologize for the "U-2 incident." Khrushchev walked out in a huff.

The show trial of Powers was a media circus that added insult to injury and cast a cloud over Eisenhower's final months in office. Powers was convicted and sentenced to ten years, but he was released in a February 1962 prisoner swap with Soviet spy Rudolf Abel.

After seeing the wreckage on television, and especially after debriefing Powers in 1962, Johnson determined what exactly had happened. "Both wings failed because of down-bending, not penetration of critical structure by shrapnel from a missile," he wrote in his memoirs. "None of the pictures showed a horizontal tail. And the right section of the stabilizer was missing. While this damage is conceivable from a crash landing, it was improbable because of the relatively undamaged condition of the vertical tail itself.

"The design of the U-2 wing is so very highly cambered that without a tail surface to balance the very high pitching moment, the aircraft immediately goes over on its back; and in severe cases the wings have broken off in down-bending. This occurred once in early testing when the pilot inadvertently extended wing flaps at high cruise speed, resulting in horizontal tail failure. This takes place in a few seconds, at great acceleration and with the fuselage generally spinning inverted. When Powers was exchanged in February 1962 for a Russian spy, I met and talked with him as soon as possible. His statements matched our conclusions."

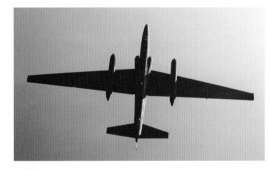

An underside view of a TR-1 tactical reconnaissance aircraft in flight in 1985. *USAF, Ken Hackman*

Captain Christopher Burns of the 1st Reconnaissance Squadron is prepped by squadron physiologists before his first solo, high-altitude U-2 flight. *USAF, Master Sergeant Dave Nolan*

Between what the Skunk Works had deduced and what Powers could add, it appeared that the missile had knocked off the right stabilizer at cruising altitude. As Johnson explains, "The airplane then, predictably, immediately went over on its back at high speed and the wings broke off in downbending. . . . With the airplane spinning badly and hanging onto the windshield for support, he tried to reach the destruct button to destroy the airplane [the CIA states that it would only have destroyed the camera]. It was timed to go off about ten seconds after pilot ejection. But he could not reach the switch. We simulated the situation and it just was not possible with the forces on his body. He had to let go."

Johnson later hired Powers at the Skunk Works. Dismissed by Lockheed after the publication of his 1971 book *Operation Overflight*, Powers went to work as a traffic reporter for radio station KGIL in Los Angeles. When he was killed in a 1977 helicopter crash while on the job, the incident spawned a round of conspiracy theories.

In the immediate aftermath of the May 1960 incident, new procedural changes required the National Security Council (NSC) to approve all CIA U-2 overflights of sensitive territory, though no more would be sanctioned of the Soviet Union or Eastern Europe.

Detachment B was closed, and all the U-2s but one were crated up and sent to the United States. Meanwhile, the Japanese government, sensitive about the "spyplanes" of Detachment C, asked that they be removed, and they were. The CIA fleet was then consolidated into Detachment G at Edwards AFB in California.

In 1961, Detachment G U-2s were twice redeployed to the Pacific, specifically to the Philippines, for some of their first operations over Laos and North Vietnam. In September 1961, during the crisis over the building of the Berlin Wall, John F. Kennedy came close to ordering a resumption, but he did not.

Meanwhile, the Joint Chiefs of Staff established the Joint Reconnaissance Center (JRC) to coordinate the separate CIA and US Air Force U-2 operations. According to Pedlow and Welzenbach, 500 missions were still being flown monthly in 1961. The targets included China, North Vietnam, and Cuba—especially before and during the ill-fated Bay of Pigs invasion in April 1961.

During 1962, CIA U-2s overflying Cuba monitored a buildup of Soviet strength on the island. On August 29, one of these flights obtained the photographs that proved a substantial effort was being made by the Soviets to establish an extensive offensive missile capability in Cuba.

After the CIA overflights confirmed the presence of Soviet medium-range ballistic missiles, responsibility for these missions shifted to the US Air Force. The first of a continuous series of flights by SAC pilots of the 4080th Strategic Reconnaissance Wing, flying out of Laughlin AFB in Texas, was made on October 14. In addition to the medium-range ballistic missiles (MRBMs), the reconnaissance also discovered numerous SAM sites. Recalling the fate of Frank Powers's U-2, the SAMs were of concern for those planning the U-2 operations—not to mention the pilots. Indeed, one of the Air Force pilots, Major Rudolf Anderson, was shot down by two SAMs and killed on October 27.

While the U-2 capability during the Cuban Missile Crisis was augmented by newly deployed reconnaissance satellites, the level of photographic detail provided by the U-2s was considered to be superior.

Brigadier General H. D. Polumbo, commander of the 380th Air Expeditionary Wing, lands his U-2R at an undisclosed location in Southwest Asia after a May 22, 2009, mission. *USAF, Senior Airman Brian Ellis*

A head-on view of a U-2 highlighting the Senior Span satellite antenna, which provides for the transmission of intelligence data in near real-time. *USAF, Master Sergeant Scott Sturkol*

A 99th Expeditionary Reconnaissance Squadron U-2R at an undisclosed base in Southwest Asia on February 23, 2010. The teardrop radome mounted above contains a Senior Span satellite communications antenna. *USAF, Master Sergeant Scott Sturkol*

A Dryden Flight
Research Center ER-2
Earth Resources aircraft
over the Sierra Nevada,
February 26, 2008.
NASA, Carla Thomas

Between 1962 and 1964, CIA U-2s deployed again to the Pacific where they conducted three dozen strategic reconnaissance missions over North and South Vietnam. After the Gulf of Tonkin incident in August 1964, however, the reconnaissance requirements in Southeast Asia shifted from strategic to tactical, and much of the responsibility for U-2 operations in the theater shifted from the CIA to the US Air Force, as a detachment of the 4080th Strategic Reconnaissance Wing deployed to South Vietnam.

However, CIA Director John McCone and the CIA covert actions oversight group known as the 303 Committee (established in 1964 by National Security Agency [NSA] Memo No. 303) continued to plan secret U-2 missions and recommend them to the president. Lyndon Johnson approved many of these, mainly on a case by case basis, from 1964 through 1966. By this time, the CIA officially noted that its total inventory of U-2s had declined to just six. These missions, especially over Cambodia, resumed in 1968, but most of the details still remain classified.

After the 1973 ceasefire, US military reconnaissance missions in Southeast Asia ended, but the Nixon administration authorized continued CIA reconnaissance overflights under Operation Scope Shield to monitor North Vietnam's adherence to the treaty provisions. In 1974, the CIA phased out U-2 operations and all of the aircraft were concentrated in the US Air Force, mainly with the 9th Strategic Reconnaissance Wing, which was formed in 1976 at Beale AFB in California.

In the late 1960s, Kelly Johnson's Skunk Works developed a new generation of U-2s. The U-2R, which would still be in service well into the twenty-first century, was substantially larger and more capable than the earlier U-2A or U-2C. It was 63 feet long, compared to just under 50 feet, and had a gross weight of 40,000 pounds, double that of its predecessors. The major difference was in the wing, the feature that had set the U-2 apart in the first place. The wingspan increased from 80 feet to 103 feet, and the wing area nearly doubled to 1,000 square feet. Increased fuel capacity, including new wing tanks, increased the range to more than 5,000 miles.

Lockheed reopened U-2R production in 1981, rolling out an improved, structurally identical variant designated as TR-1 (for Tactical Reconnaissance). These new versions took their places alongside existing U-2Rs. By this time, the aircraft was routinely known as the "Dragon Lady," a name that dated back to the original transfer of CIA U-2s to the Air Force in 1956. In 1992, all the existing U-2Rs and TR-1s were consolidated under the U-2R designation. Later in the decade, U-2Rs were reengined with General Electric F118 turbofans and redesignated U-2Ss.

In the meantime, two of the new production TR-1s were transferred to NASA under the designation ER-2 (for Earth Resources), bringing the story of the aircraft back to the cover story that was first used in 1960 when the aircraft was inconveniently revealed to the world by Nikita Khrushchev.

The Dragon Lady had long since departed the mysterious Groom Lake of her birth, but inside Area 51, many things had been happening during the decades that the U-2 had been operational.

A portrait of a U-2 pilot (or in this case, an ER-2 pilot) at work. NASA's Tom Ryan snapped this self-portrait while flying a Multiple Altimeter Beam Experimental Lidar (MABEL) laser altimeter mission on April 5, 2011. *NASA, Tom Ryan*

CHAPTER 6
IMAGINING THE ARCHANGELS

AT ITS INCEPTION, the U-2 had been a groundbreaking—or ceiling breaking—innovation in terms of its altitude capabilities. However, as it entered service, shortcomings came into focus. Its relatively slow, subsonic speed contributed to its vulnerability, and after a time, advances in Soviet missile and interceptor technology made its once-unchallenged service ceiling seem not so remote.

Even before the Powers debacle in 1960, efforts were being made on both the government and industry sides toward a higher and faster reconnaissance aircraft to supersede the U-2. After the disappointing results of the Project Rainbow "dirty bird" experiments, an interest grew in having radar-obscuring technologies built into a new aircraft design, rather than retrofitting the plane.

"We knew we needed more altitude and, especially, more speed," wrote Kelly Johnson in his memoirs. "Vulnerability studies led us to the decision that the next airplane should operate at altitudes well over 80,000 to 85,000 feet, fly at speeds well over Mach 3, and be able to out-maneuver any SA-2 missile the Russians might develop. It had to be stable enough in flight to take a good photograph from altitudes above 90,000 feet. It had to retain the characteristics of the U-2—be able to photograph very, very small targets on the ground—while flying four to five times as fast. We wanted it to have global range—with multiple midair refueling from the KC-135 aerial tankers. The aircraft also should present an extremely low radar cross section—be very difficult to detect."

The times called for out-of-the-box innovation, but they were remarkable times. In aerospace, it was a golden age, an era when even the sky was not the limit—especially to the imagination. At the beginning of the 1950s, nobody had yet flown at twice the speed of sound; by the end of 1963, the X-15 had topped Mach 6. In the beginning of the 1950s, space flight was science fiction; by 1963, astronauts and cosmonauts were in space and anticipating a near-term trip to the moon.

It was an era when technology promised an unlimited future. Nuclear propulsion for ships had become a reality, and many in the aerospace industry, especially at Lockheed and the Convair Division of Genera Dynamics, were anticipating nuclear propulsion for aircraft. Even as such projects as Pluto and NERVA were ongoing at the Nevada Test Site,

Known to the CIA as Oxcart and to Lockheed as the Archangel, the fastest air-breathing airplane in the world in the 1960s was developed in complete secrecy by the Lockheed Skunk Works. Here, Article 133, the final Lockheed A-12, takes shape at Plant B-6 in Burbank. *Lockheed*

the AEC had established the National Reactor Testing Station in a remote location in the high desert country of southern Idaho. It was here, inside lead-lined hangars, that a potential reactor for an airplane was to have been developed.

In 1955, Convair first flew a functioning R-1 nuclear reactor aboard a modified NB-36H. Meanwhile Kelly Johnson's Skunk Works team was pushing the edge of the innovation envelope with the Lockheed Model CL-400. Being developed under the ARDC code name Suntan, this aircraft was intended to achieve Mach 3 speeds with engines fueled by liquid hydrogen. While liquid hydrogen never matured into a practical

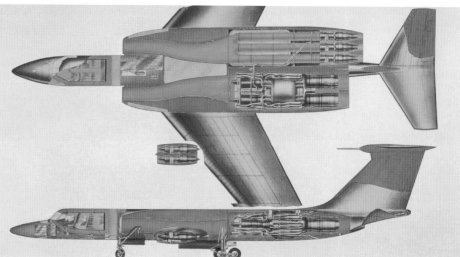

Above: The A-12 configuration was "pole tested" at Groom Lake to evaluate its radar cross section (RCS). The blending of the fuselage into the wings helped considerably in reducing the RCS. *Author's collection*

Right: During the early 1950s, Lockheed was one of several companies that conducted engineering studies of various configurations for a nuclear-powered aircraft. *Lockheed*

propulsion technology, and Suntan was finally cancelled in 1959, the program did allow Johnson's engineers to explore numerous airframe configurations that would be compatible with speeds of Mach 3 and above.

When the CIA came calling in late 1957, both Convair and Lockheed were developing the necessary expertise to entertain thoughts of creating an aircraft capable of cruising at speeds beyond which any aircraft had yet cruised.

As noted previously, Richard Bissell, the special assistant to DCI Allen Dulles who was the CIA's U-2 program manager had reckoned early on that the U-2's service life would be finite and was anxious for something new—and something unimaginably advanced. He had initiated Project Gusto, a highly classified search for the new generation Mach 3 spyplane.

At Convair, Vincent "Vinko" Dolson and Bob Whidmer created a small, dart-shaped, ramjet-powered aircraft theoretically capable of speeds above Mach 4 and operations at an altitude of 125,000 feet. This aircraft, initially called "Super Hustler," was to have been air-launched by a stretched Hustler designated as B-58B. In an effort to reduce the RCS of the aircraft, the Convair engineers designed the nose and leading edges of the aircraft to be made of heat-resistant ceramic Pyroceram. Because this would supposedly render it "invisible" to radar, the aircraft became known as the FISH (for First Invisible Super Hustler).

The complexity of requiring a still unproduced "mothership" complicated the process considerably.

As Kelly Johnson later commented, "Convair proposed a ramjet-powered Mach 4 aircraft, which also had to be carried aloft by another vehicle and launched at supersonic speeds where the ramjet power could take over. Unfortunately the launch vehicle was the B-58, which could not attain supersonic speed with the bird in place. Even if it could, the survivability of the piloted vehicle was in question because of probable ramjet blowout in maneuvers. The total flight time for the Marquardt ramjet at the time was not over seven hours, obtained mainly on the ramjet test vehicle for the Boeing Bomarc missile."

Coincidentally, this test vehicle, the X-7, had been created by the Skunk Works.

For all of these reasons, the CIA turned down the FISH concept, as well as the B-58B variant, and neither was built. Undaunted Convair went back to the drawing board, developing their Kingfish vehicle, which would take off conventionally from a runway. Like the FISH, Kingfish used Pyroceram to achieve a low RCS, and like the FISH and the B-58, it was a delta-winged craft. It was 76.5 feet long, 20 feet shorter than the B-58, but its wingspan of 60 feet was 3 feet greater.

At the Skunk Works, Kelly Johnson's design had its roots in the CL-400 design studies for Project Suntan. As the U-2 had been known internally as the Angel, it was natural that the Skunk Works should call their new creation the Archangel.

Johnson's Skunk Works team included Henry Combs and Ray McHenry on structures, Merv Heal and Lorne Cass for weights and loads, and Dick Cantrell and Dick Fuller on aerodynamics, while Dave Robertson designed the fuel system. The cockpit was designed by Dan Zuck with input from Lockheed test pilot Lou Schalk, whom Kelly Johnson had hand-picked to fly the Archangel test-flight program. Dave Campbell and Ben Rich were involved with the propulsion system.

Kelly Johnson's September 1958 sketch of the notional third Archangel (A-3) was an early incremental step on the road that led to the A-12. *CIA*

With a master's degree in aeronautical engineering from UCLA, Rich had been with the Skunk Works since the F-104 program and had been part of the team that built the U-2. He was now the chief aerodynamicist and program manager for the Archangel project. In 1975, he would become Johnson's successor as head of the Skunk Works.

There would be a dozen Archangel designs, numbered A-1 through A-12. The early Archangels looked a bit like scaled-up Starfighters, while others were shaped like arrowheads. Some had wingtip engines, but others blended the engine nacelles into the middle of the wings.

It was an era of new design challenges beyond the experience of most engineers. "All the fundamentals of building a conventional airplane were suddenly obsolete," Ben Rich recalled in his memoirs. "Even the standard aluminum airframe was now useless. Aluminum lost its strength at 300 degrees Fahrenheit, which for our Mach 3 airplane was barely breaking a sweat. At the nose the heat would be 800 degrees—hotter than a soldering iron—1,200 degrees on the engine cowlings, and 620 degrees on the cockpit windshield, which was hot enough to melt lead. About the only material capable of sustaining that kind of ferocious heat was stainless steel."

However, Kelly Johnson decided on titanium because it was a material with which the Skunk Works had been working for a decade.

The final variant of the Archangel lineage, the A-12, had midwing engines and a fuselage that blended smoothly into the wings. Notable in the blended design were a chine, or wing leading edge extension, on each side of the forward fuselage. These curved up from the wing, flowing into the nose, providing useful additional lift at supersonic speeds. As Johnson puts it, "The shape of the airplane itself was determined after a great many wind-tunnel and other tests. The result, head on, looks like a snake swallowing three mice."

Ben Rich writes that the A-12 "weighed 96,000 pounds without fuel—it's light to maximize fuel consumption and minimize cost—and was 108 feet [actually 101 feet 3 inches; the later SR-71 was nearly 108 feet] long with an extremely thin double delta wing attached at mid-fuselage. The wing edge was designed so razor-thin that it could actually cut a mechanic's hand."

Below left: On April 21, 1958, Kelly Johnson wrote in his official diary that he "drew up the first Archangel proposal for a Mach 3 cruise airplane having a 4,000 nautical mile range at 90,000 to 95,000 feet." This was the A-1. *CIA*

Below right: Convair's First Invisible Super Hustler (FISH) was a contender in 1958 for the CIA's high-speed, radar-evading reconnaissance aircraft. *CIA*

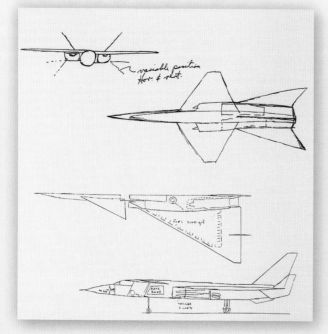

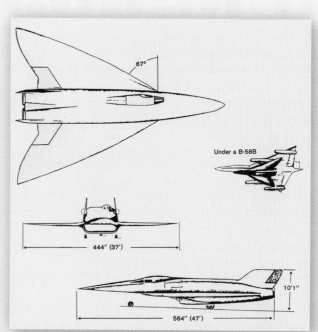

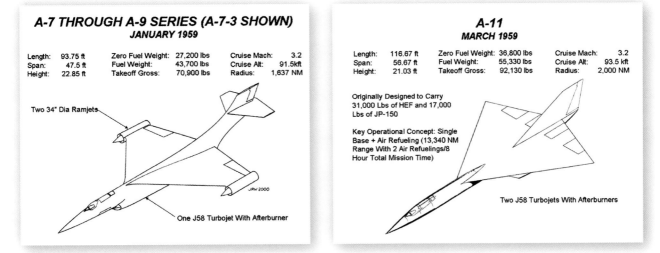

A-7 THROUGH A-9 SERIES (A-7-3 SHOWN)
JANUARY 1959

Length:	93.75 ft	Zero Fuel Weight:	27,200 lbs	Cruise Mach:	3.2
Span:	47.5 ft	Fuel Weight:	43,700 lbs	Cruise Alt:	91.5 kft
Height:	22.85 ft	Takeoff Gross:	70,900 lbs	Radius:	1,637 NM

Two 34" Dia Ramjets

JRW 2000

One J58 Turbojet With Afterburner

A-11
MARCH 1959

Length:	116.67 ft	Zero Fuel Weight:	36,800 lbs	Cruise Mach:	3.2
Span:	56.67 ft	Fuel Weight:	55,330 lbs	Cruise Alt:	93.5 kft
Height:	21.03 ft	Takeoff Gross:	92,130 lbs	Radius:	2,000 NM

Originally Designed to Carry
31,000 Lbs of HEF and 17,000
Lbs of JP-150

Key Operational Concept: Single
Base + Air Refueling (13,340 NM
Range With 2 Air Refuelings/8
Hour Total Mission Time)

Two J58 Turbojets With Afterburners

The A-12 had a wing span of 55 feet 7 inches, a wing area of 1,795 square feet, and a tail that stood 18.5 feet high. It had a gross weight of 117,000 pounds. Though the A-12 had a stated service ceiling of 90,000 feet, it would officially top 96,000.

The engines specified for both the Archangel and Kingfish projects were to be a pair of Pratt and Whitney J58 turbojets, which delivered a maximum thrust of 34,000 pounds. Originally developed for a high-speed variant of the US Navy's Vought F8U fighter that was never built, the J58 was designed to operate for extended periods on afterburner, providing sustained Mach 3 cruise capabilities. Like the aircraft themselves, the J58 embodied expensive and complex leading-edge technology. Because of the high-altitude and high-temperature environment in which the aircraft would operate, the J58 operated on JP-7 fuel, which had such extremely low volatility that it required a chemical igniter called triethylborane to be injected into the engine to initiate combustion. As this author has seen personally, a lighted match dropped into JP-7 will not ignite it.

"Our engines were the only items off the shelf," Ben Rich explained. "[The J58], which would need major modifications for our purposes, had already undergone about seven hundred hours of testing before the government cut off its funding. Each of these engines was Godzilla, producing the total output of the *Queen Mary*'s four huge turbines, which churned out 160,000 shaft horsepower. Using afterburners at Mach 3, the exhaust-gas temperatures would reach an incredible 3,400 degrees. This propulsion system would not only be the most powerful air-breathing engine ever devised but also the first ever to fly continuously on its afterburners, using about eight thousand gallons of fuel an hour."

Meanwhile, the Convair and Lockheed proposals were among those that the CIA received and considered under Project Gusto.

"Some of the other entries were interesting," Kelly Johnson observed, noting one of the more unusual. "A Navy in-house concept proposed an inflatable rubber vehicle which could be carried to altitude by a balloon,

Above left: Wingtip engine placement and a delta wing figured prominently in the later Archangel design studies. *CIA*

Above right: With its single tail and underwing engines, the penultimate Archangel was still a far cry from the final A-12 configuration. *CIA*

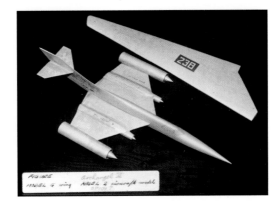

This small scale model of the second Archangel (A-2) included the blended-nacelle, midwing engine placement that would later be used in the final configuration. *Lockheed*

then boosted by rocket power to a speed where its own ramjets could take over. This rapidly was demonstrated to be unfeasible. The balloon would have had to be a mile in diameter to lift the unit, which itself had a proposed wing area of one-seventh of an acre."

With the Convair Kingfish and the Lockheed A-12 as the leading candidates, the final presentation came on August 20, 1959. The proposals were evaluated by a panel that included the US Air Force and the DOD, as well as the CIA. It took them just a week to make their decision.

Richard Bissell met with Johnson privately on August 28 to tell him that Lockheed would get the contract. As Johnson wrote in his official project log, "Saw Mr. Bissell alone. He told me that we had the project and that Convair was out of the picture. The agency accepts our conditions that, our method of doing business will be identical to that of the U-2. Mr. Bissell agreed very firmly to this latter condition and said that unless it was done this way he wanted nothing to do with the project either."

In addition to the design itself, other factors in Lockheed's favor were its proven performance during the U-2 program, including the fact that the U-2 had stayed below budget while Convair's B-58 program had suffered under cost overruns.

As a project name, the CIA picked "Oxcart." Ben Rich called this name "an oxymoron to end all: at Mach 3, our spy plane would zip across the skies faster than a high-velocity rifle bullet."

Initially Lockheed received funding for five A-12 aircraft to be built over two years. The quoted price was $96.6 million. Though the actual contract was not issued until early 1960, the Skunk Works began work on the full-scale A-12 mock-up three days after Johnson met with Bissell. This work and the A-12 manufacturing took place in the factories of Lockheed Plant B-6 complex in Burbank, which had seen the production of thousands of P-38 fighter planes during World War II.

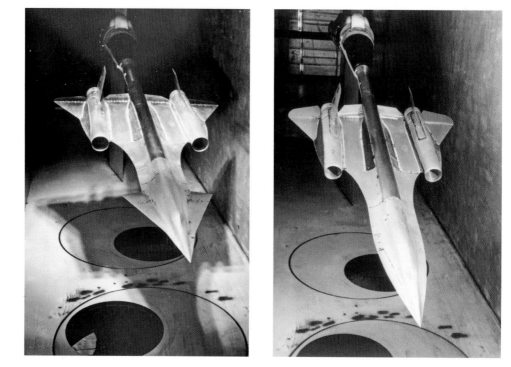

Left: This wind tunnel model of an early A-12 variant had the fuselage-wing blending of the final but featured a sweeping double-delta configuration. *Lockheed*

Right: The final A-12 configuration is seen in this wind tunnel model, circa 1959. *Lockheed*

Left: Manufacturing the most advanced airplanes in the world in the secret depths of Lockheed's Plant B-6, circa 1963. *Lockheed*

Below: A map of the Lockheed Plant B-6 complex in Burbank. Most A-12 and later SR-71 production took place in Buildings 309 and 310, just inside the main gate at 2901 Hollywood Way. *Burbank-Glendale-Pasadena Airport Authority*

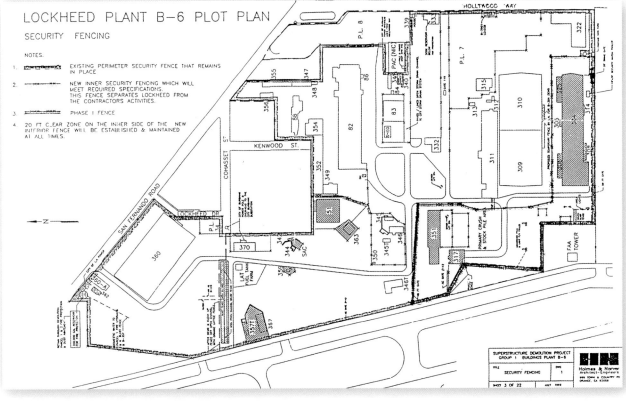

LOCKHEED PLANT B-6 PLOT PLAN

SECURITY FENCING

NOTES:

1. EXISTING PERIMETER SECURITY FENCE THAT REMAINS IN PLACE

2. NEW INNER SECURITY FENCING WHICH WILL MEET REQUIRED SPECIFICATIONS. THIS FENCE SEPARATES LOCKHEED FROM THE CONTRACTORS ACTIVITIES.

3. PHASE I FENCE

4. 20 FT CLEAR ZONE ON THE INNER SIDE OF THE NEW INTERIOR FENCE WILL BE ESTABLISHED & MAINTAINED AT ALL TIMES.

Throughout 1960, the Oxcart program was consumed with structural and aerodynamic testing and with working through the myriad of bugs that can be expected in the construction of something unlike anything that had ever been made. As the A-12 prototype was under construction, entire sections were completed and tested. As Johnson recalls, "Before we got into high gear on production, we thought it advisable to build several test samples of the most complex sections: the nose and the basic wing structure. The first wing section was a catastrophe. When we put it in a 'hot box' to simulate high inflight temperatures, it wrinkled up like an old dishrag. The solution was to divorce the skin panels from the wing spars in each direction and put corrugations and dimples in the skin—the wing surface. When the titanium got really hot, the corrugations merely deepened."

In fact, the Skunk Works had to invent a very large press that would shape titanium under very high temperatures up to 1,500 degrees Fahrenheit.

"The nose section of the airplane presented other problems," Johnson observed. "We put it in the hot box to study cooling requirements for the pilot and the gear. We produced 6,000 parts, and of them fewer than ten percent were any good. The material was so brittle that if you dropped a piece on the floor it would shatter. Obviously, we were doing something wrong."

When the first A-12, Article 121, was nearly completed in January 1962, it was placed into gigantic, custom-built wooden crates for shipment to Groom Lake. *Lockheed*

Johnson queried Titanium Metals Corporation (TMC) on this problem, but they didn't know. He ordered the entire titanium processing system to be trashed and replaced it with the same methods TMC was using in their factory to manufacture the original sheets and forgings.

"After the initial shambles on the nose segment heat treat tests, we put into

Article 121, the first A-12, departed Burbank at 2:30 a.m. on the drizzly morning of February 26, 1962 for its trip to Groom Lake. Here, the convoy of 18-wheelers, under California Highway Patrol escort, pulls over to let traffic pass. *Terry Panopalis collection*

Having reached Groom Lake, A-12, Article 121, is seen here on the flight line behind temporary revetments during fuel system and engine checks. A leak was discovered, requiring top-mounted external tanks for the testing. *Lockheed*

effect a quality-control program that I believe was and is unequaled anywhere. For every ten parts manufactured, we made three sample parts. These would be heat-treated and otherwise tested before any of the others of the batch would be put in storage for future use. . . . We could trace back to the mill and know the direction of the sheet rolling, and whether the part was cut with or against the grain. Before we would do all the expensive machining to cut landing gears from the huge heavy extrusions we would cut twelve samples, and unless every one met the test we devised for them, we would not use that extrusion to make a landing gear."

The original delivery date for the first A-12 slipped from August 1961 to the end of the year, but by January 1962, it was ready for its first flight. Because of the extremely high security surrounding this blackest of black airplanes, it certainly could not be tested at Burbank—nor even at Edwards AFB. There was only one place to go, and that was back to the rookery of the Angels at Groom Lake.

Indeed, work was already underway inside Area 51 to make it ready for the Archangel. The runway was being lengthened from 5,000 to 8,500 feet, and the hamlet of mobile homes that once housed the U-2 family was being replaced with permanent frame buildings. Hangars acquired from the US Navy were disassembled and brought in. The CIA arranged secretly with the FAA to expand the restricted air space over Area 51.

In the official CIA history of Oxcart, Gregory Pedlow and Donald Welzenbach note that "air controllers were warned not to mention the craft on radio but to submit written reports of sightings or radar trackings."

This naturally provided more grist for the conspiracy theorists and those reporting unidentified objects to Project Blue Book.

Once finished, the first A-12, known as "Article 121" and given the US Air Force tail number 60-6924, was disassembled and packed in large crates. In the predawn darkness of February 26, 1962, it left Burbank by truck. By early afternoon, under the watchful eye of Dorsey Kammerer, the prototype of the ultimate twentieth century spyplane was inside Area 51.

CHAPTER 7
ARCHANGELS OVER AREA 51

A NERVOUS KELLY JOHNSON landed at Groom Lake on April 24, 1962. The following day, an airplane like none before it would make its debut flight. This Archangel was more than just another of Johnson's babies—it was the capstone of his amazing career.

Johnson was as nervous over the first flight as anyone in his position is on the eve of any first flight. He was also nervous because of numerous problems that had arisen after the A-12 had arrived inside the gates of Area 51. Not the least of these were issues involving fuel having softened the sealant in fuel tank seams.

His concerns also naturally included pleasing his customer, although Johnson typically liked not simply to please his customers, but to amaze them.

In the months since work on the A-12 prototype, Article 121, had begun, some big changes had occurred inside the customer's home office in Langley, Virginia. Though none of these would effect the Oxcart program immediately, new faces would oversee the eventual deployment of the Oxcart. Originally retained by the incoming Kennedy administration in January 1961, both Allen Dulles and Richard Bissell had now departed the CIA. Dulles was succeeded as DCI by John McCone. Herbert "Pete" Scoville succeeded Bissell as the CIA's DPD was transferred from the directorate of plans to the directorate of research. Colonel Leo Geary, the US Air Force project manager for the U-2, remained in that role for the Oxcart program.

As with the Angel before it, the first actual flight of the Archangel was deemed unofficial. It was just a hop above a taxi test, with Lou Schalk taking Article 121 up just 200 feet off the bed of Groom Lake and flying only about a mile and a half. Some problems with center of gravity occurred because there was no fuel in the forward tanks, but no other major issues presented themselves. The next day, April 26, Schalk took the aircraft up to 30,000 feet and a top speed of 390 mph, touching down with a thumbs up after 59 minutes. The early flights were made using off-the-shelf J75 engines. The J58 would not be ready until October.

The first flight with the customer present took place on April 30. Pete Scoville magnanimously invited the retired Richard Bissell to join him on the flight line. In his project log, Kelly Johnson wrote, "I was very happy to have Dick see this flight, with all he has contributed to the program."

The first supersonic flight of Article 121 took place on May 4, and the second A-12, Article 122, arrived in Area 51 in June; it was diverted to pole tests of its RCS. By December 19, all five of the initial batch of Archangels—including one twin-seat A-12B trainer—had reached the Nevada desert, though they would not all be flying until the middle of 1963. Increased orders brought the total Archangel population at Groom Lake to nine by the end of 1963.

The first flight test with two J58 engines took place on January 13, 1963, but the performance was not as expected. Pratt & Whitney spent the next several months working out the bugs. Additional test pilots, including Bill Park, Jim Eastham, and Bob Gilliland, joined the test program as the improved J58s arrived, and by the end of 1963, the aircraft were flying routinely above Mach 3.

Meanwhile DCI McCone found himself defending Oxcart to the budget-conscious new Secretary of Defense Robert McNamara, whose axe was famously falling on such far-reaching programs as the North American XB-70 Valkyrie Mach 3 intercontinental bomber as well as the Boeing X-20 DynaSoar, which would have given the United States a spaceplane capability more than a dozen years before the space shuttle.

An A-12 Oxcart taxis past a US Air Force F-101B chase plane on the paved runway at Groom Lake. Aircraft of the A-12 fleet spent most of their service lives flying out of this facility. *Terry Panopalis collection*

Ben Rich later recalled that the first time Kelly Johnson met McNamara: "He found him haughty and cold."

"That guy will never buy into a project that he hasn't thought up himself," Johnson told staff at a meeting soon after. "He's petty, the kind who will throw out any project begun under Eisenhower. He just doesn't believe that anyone else has his brains and he'd love to stick it to an old-timer like me just to show the entire aerospace industry who's boss."

In July 1962, McNamara argued that the Oxcart would never be used and that it should be eliminated. However, according to Gregory Pedlow and Donald Welzenbach, McCone disagreed with McNamara, telling the secretary that he had every intention of flying the aircraft over the Soviet Union. At the time, that seemed to be probable. There was growing concern about the deployment of antiballistic missile systems within the Soviet Union, and the program aimed at a first generation of Corona surveillance satellites was running into problems.

McCone told President Kennedy that the Oxcart was the only viable alternative. After the Cuban Missile Crisis in October 1962, there seemed to be no doubt that the United States needed a robust overhead reconnaissance program.

Since the very beginning of the Oxcart program, the US Air Force had been expressing an interest of its own in the remarkable Archangel. Like any good salesman, Kelly Johnson readily encouraged them by proposing a fighter-interceptor, which he called the AF-12.

Back in 1955, the US Air Force had begun development of a Mach 3 interceptor, the North American Aviation F-108 Rapier. This project had been cancelled in 1956 without

Test pilot Lou Schalk is congratulated at Groom Lake after making the first "official" A-12 flight on April 30, 1962. He shakes the hand of a man identified as Richard Bissell, who had left the CIA but was invited to the flight by his successor, Pete Scoville. The hatless man with his hands clasped behind him is Najeeb Halaby, the administrator of the FAA and future CEO of Pan American Airways. *Terry Panopalis collection*

a completed aircraft, but interest in a Mach 3 interceptor remained. In October 1960, the air force commissioned Lockheed to go ahead with the aircraft that would become the YF-12A.

The YF-12 was 101 feet 8 inches long, just 5 inches longer than the A-12, but it had the same wing span (55 feet 7 inches), the same wing area (1,795 square feet), and a tail height of 18.5 feet. The gross weight was 124,000 pounds. The stated service ceiling of 75,000 feet was more like a low estimate.

The YF-12A was similar in structure and appearance to the A-12, except for a substantially redesigned forward fuselage. This included a second cockpit for the weapons system operator, as well as accommodation for the large and complex Hughes AN/ASG-18 radar fire control system. The redesigned nose bore distinctively truncated chines that make the YF-12A instantly distinguishable from an A-12.

The first American pulse-doppler radar with look–down shoot down capability, the AN/ASG-18 had originally been developed for the canceled F-108. The YF-12A also inherited the Hughes AIM-47 (originally GAR-9) Falcon air intercept missile, which had been intended for the F-108. The AIM-47 was carried internally inside a bay that on the A-12 carried reconnaissance cameras.

The debut flight of the first YF-12A (tail number 60-6934, Article 1001), with Jim Eastham at the controls, occurred at Groom Lake on August 7, 1963.

By now, the ongoing A-12 flight testing over the Nellis Range was attracting more than the desirable amount of attention. There was a great deal of talk within the aerospace media that something very fast was being tested out there. In April 1963, to address this problem, Pete Scoville proposed that the US Air Force should announce the YF-12A as an A-12 cover story and say that it was already testing its new high-speed interceptor.

As Pedlow and Welzenbach point out in the official history of Oxcart, two schools of thought existed on the topic within the CIA. One side opposed any disclosure of the program whatsoever, while Scoville and others argued that there were already rumors and sightings, and it would be necessary for Oxcart to be "surfaced" by a cover story.

This late 1963 family portrait of A-12s includes two YF-12As parked at the far end. Second in line is Article 124, the Titanium Goose, the only A-12B two-seat trainer. It has been said that this group photo included all extant A-12s, which would date it to late 1963. *Lockheed*

The fifth Oxcart, Article 125, made its debut flight at Groom Lake in January 1963 with Bill Park at the controls. It crashed in December 1965 due to a malfunction traceable to improper wiring. *Terry Panopalis collection*

President Lyndon Johnson would be cast in the central role in the surfacing of Oxcart. He took office upon the assassination of John F. Kennedy on November 22, 1963, and was briefed on Oxcart by DCI McCone one week later. The surfacing came on February 29, 1964.

"The United States has successfully developed an advanced experimental jet aircraft, the A-11, which has been tested in sustained flight at more than 2,000 miles an hour, and at altitudes in excess of 70,000 feet," Johnson told reporters in the International Treaty Room at the Department of State. "The performance of the A-11 far exceeds that of any other aircraft in the world today. The development of this aircraft has been made possible by major advances in aircraft technology of great significance for both military and commercial application. Several A-11 aircraft are now being flight tested at Edwards Air Force Base in California.

"The existence of this program is being disclosed today to permit the orderly exploitation of this advanced technology in our military and commercial programs. This advanced experimental aircraft, capable of high speed and high altitude, and long-range performance at thousands of miles, constitutes a technological accomplishment that will facilitate the achievement of a number of important military and commercial requirements.

"In view of the continuing importance of these developments to our national security, the detailed performance of the A-11 will remain strictly classified and all individuals associated with the program have been directed to refrain from making any further disclosure concerning this program."

He took no questions on the subject.

There was, of course, no such thing as an A-11. That particular Archangel had never been built. As Pedlow and Welzenbach write, "Johnson's use of the A-11 designator at the press conference has sometimes been called an error, but Kelly Johnson wrote the President's press release and chose that designator for security reasons because it referred to an earlier version of the aircraft that lacked the radar-defeating modifications of the A-12."

Nor was this "advanced experimental jet aircraft" being tested at Edwards AFB—although it did make a cameo appearance there in order to maintain the secrecy of the Groom Lake facility. Then, Area 51 did not exist.

By this time, both the A-12 and the YF-12A were well into their test flight program, and the US Air Force reconnaissance variant was moving forward at the Skunk Works facility in Burbank.

Early in the A-12 program, the US Air Force had noncommittally expressed an interest in a reconnaissance variant of its own, and there had also been talk of adapting the A-12 airframe as a Mach 3 bomber. In early 1962, the Skunk Works worked up mock-ups of the two options: an R-12 reconnaissance aircraft and an RS-12 "reconnaissance-strike" variant. Both aircraft were longer and heavier than the A-12 and carried a two-man crew.

In early 1963, under the code name Senior Crown, the US Air Force gave Lockheed the green light to build six R-12s, with the possibility of twenty-five more, for SAC. The "senior" prefix in two-word code names is believed to indicate programs that are of direct interest to US Air Force Headquarters.

As Ben Rich recalls, the way the US Air Force came to buy its own reconnaissance Archangel was almost comical. It came during a visit to the Skunk Works by General Curtis LeMay, the former SAC commander who became Air Force Chief of Staff in 1961. The subject on the table had been the YF-12A.

"We'll buy your interceptors," LeMay told Johnson. "I don't have a number yet but I'll get back to you soon."

In fact, the US Air Force initially ordered eighteen YF-12As, but the order was later trimmed to just three.

"What about the reconnaissance aircraft we built for the agency?" Johnson then asked. "Can't the Air Force use any?"

"You mean, we haven't ordered any?" LeMay replied, flabbergasted.

As Rich recalls, LeMay "wrote a note to himself and promised Kelly he would forward an Air Force contract for the two-seater version of the spyplane within a few

The Titanium Goose, the sole A-12B trainer, taxis on the Watertown landing strip at Groom Lake. It was also the A-12 with the greatest flight time, accumulating 1,076.4 hours in 614 flights, more than double any other A-12. *Terry Panopalis collection*

A-12s sometimes wore bogus US Air Force markings to obscure the fact that they were CIA aircraft. *Lockheed*

weeks. The very next day, Kelly was tipped off by a colonel on LeMay's staff that on the trip back to Washington aboard his jet, LeMay revised his thinking rather sharply and ordered his staff to develop a proposal for building ten Blackbird interceptors and ten tactical bombers a month!"

Though this apocryphal prediction never came to pass, the aircraft intended for reconnaissance was ordered, and it was assigned a number (71) in the US Air Force pre-1962 "B-for-bomber" lineage. However, the "B" prefix was never assigned.

In the meantime, in an effort to restart the cancelled North American XB-70 bomber program, the redesignation as RS-70 had been proposed as a way to underscore its multimission capabilities. Therefore, the Skunk Works aircraft was briefly known as the RS-71 before it became the SR-71. By some accounts, General LeMay favored the "SR-for-strategic-reconnaissance" prefix. By other accounts, President Lyndon Johnson accidentally transposed the letters when he announced the existence of the 71.

The reason that Johnson lifted the veil on a second secret aircraft in the space of five months is almost universally believed to have been motivated by politics. It was an election year, and he was locked in a contentious reelection battle with challenger Senator Barry Goldwater. The latter was campaigning on the issue that the United States was falling behind the Soviet Union in the development of leading edge military technology, and Johnson needed to play a trump card.

On July 24, 1964, five months after he had revealed the "A-11" (A-12), Johnson announced to reporters "the successful development of a major new strategic manned aircraft system, which will be employed by the Strategic Air Command. This system employs the new SR-71 aircraft, and provides a long-range, advanced strategic reconnaissance plane for military use, capable of worldwide reconnaissance for military operations. The Joint Chiefs of Staff, when reviewing the RS-70, emphasized the

importance of the strategic reconnaissance mission. The SR-71 aircraft reconnaissance system is the most advanced in the world. The aircraft will fly at more than three times the speed of sound. It will operate at altitudes in excess of 80,000 feet. It will use the most advanced observation equipment of all kinds in the world.

"The aircraft will provide the strategic forces of the United States with an outstanding long-range reconnaissance capability. The system will be used during periods of military hostilities and in other situations in which the United States military forces may be confronting foreign military forces."

Twice, and with no apparent strategic justification, Johnson had provided the world, and America's antagonists, with details of the country's most advanced strategic aircraft. This had been done before any of the variants had become operational.

He had not, however, disclosed the existence of Area 51.

The SR-71 had the same wingspan and wing area as the A-12 and YF-12A (55 feet 7 inches and 1,795 square feet), but it was about 6 feet longer at 107 feet 5 inches because of its second cockpit. As Lyndon Johnson suggested, its operational altitude was above 80,000 feet—and is widely believed to have been above 100,000 feet. As the Senior Crown program began, Lockheed moved the SR-71 final assembly from Burbank to the more secure Air Force Plant 42 at Palmdale, near Edwards AFB, in a factory previously utilized by North American Aviation. The first SR-71 would make its debut flight at Palmdale on December 22, 1964.

Meanwhile, inside Area 51, the A-12 and YF-12A were still in flight test. The second and third YF-12As made their first flights in November 1963 and March 1964 respectively, and Jim Eastham first successfully sustained speeds of Mach 3.2 in the YF-12A in January 1965. The first release of an inert AIM-47 by a YF-12A came on April 16, 1964, and the first full-blown missile launch, on March 18, 1965, achieved a hit on a Q-2 target drone

This 1960s photograph of the Groom Lake complex shows the four large hangars and the housing area. The main runway ended at the edge of the dry lakebed during the U-2 era but was lengthened so that the A-12 and later aircraft could land and takeoff on a paved surface. *Tony Landis collection*

at a range of 36.2 miles. Results in such simulated intercepts were, however, disappointing. Seven tests, conducted through September 1966, resulted in only two hits. Most of these were conducted over the Pacific Ocean, and none were in Groom Lake air space.

Within the DOD, the idea of a new Mach 3 interceptor, a decade after the cancellation of the F-108, never gained much traction. With the shift of priority from manned bombers to ICBMs, the Soviet Union seemed unlikely to develop a Mach 3 bomber for such an aircraft to intercept. The YF-12A program ended, quietly by what the outside world would perceive but not so quietly if you were there. The first of the three aircraft was lost in a hard landing in August 1966, and the third in a nonfatal crash in June 1971. The remaining YF-12A was consigned to storage in 1967 but loaned to NASA in 1969 for Mach 3 testing.

The A-12 test program continued through most of 1965, logging considerable Mach 3–plus time despite the loss of three of the fifteen aircraft—and one pilot—between May 1963 and December 1965. Much to Kelly Johnson's frustration, the teething troubles had more to do with manufacturing errors than design flaws.

The third and last YF-12A is pictured in flight over the Nellis Range. The nose was substantially redesigned from that of the A-12, with distinctively truncated chines. This modification was necessary to accommodate the large and complex Hughes AN/ASG-18 radar fire control system. *Lockheed*

The errors, in turn, contributed to delays in qualifying CIA pilots to take their places alongside the Lockheed test pilots. In the fall of 1964, when the CIA wanted to conduct overflights of Cuba with its Oxcarts, it had no pilots. Kelly Johnson offered to fly these missions with Lockheed pilots, but the record is unclear as to whether or not these overflights actually took place.

The first YF-12A made its debut flight from Groom Lake in August 1963 but was lost in August 1965 in a hard emergency landing at Edwards AFB after an inflight fire. *Terry Panopalis collection*

According to CIA records quoted by Gregory Pedlow and Donald Welzenbach, the "final validation flights for Oxcart [operational] deployment" concluded on November 20, 1965. On that date, the CIA formally activated its own 1129th Special Activities Squadron (SAS), although it would be eighteen months before they made that operational deployment. In his log, Johnson notes that as of May 12, 1966, "there is still no go-ahead for deployment, although it seems fairly optimistic. The airplanes are ready to go."

As they awaited orders, the CIA was still evaluating the capabilities of its A-12 fleet, which was still operating out of Groom Lake. As noted by Pedlow and Welzenbach, the files of the CIA OSA contain the details of many remarkable flights, including one in which Lockheed test pilot Bill Park took off from Groom Lake on a nonstop flight that took him to Duluth, Minnesota, and then south to Tampa by way of Atlanta. He next turned northwest to Portland, Oregon, turned about and flew to Knoxville by way of Denver and St. Louis, and then returned to Groom Lake. He logged more than 10,000 miles—in just six hours.

Park had proven that the Archangel could do what no other aircraft could do, but it would be more than a quarter of a century—and many record flights by cousin SR-71—before this particular accomplishment would be made public.

The second YF-12A (top) is shown here with the second SR-71 (61-7951), the latter having been marked with a bogus tail number (60-6937) in the YF-12A sequence. As a cover story, it was called the "YF-12C." Beneath the belly of the aircraft is a heat transfer fixture that exposed experiments to rapid temperature rises during sustained supersonic flight. *NASA*

To support long distance flights by the CIA A–12s, as well as their own YF–12As and SR–71s, the US Air Force converted a number of Boeing KC–135A Stratotanker aerial refueling aircraft to offload JP-7 jet fuel rather than standard JP-4 and redesignated these planes as KC–135Qs. Whenever and wherever the Lockheed Mach 3 jets were deployed, they were accompanied by KC–135Q tankers.

In 1965, Operation Upwind, also called Project Scope Logic, envisioned the use of A–12s based in Britain to conduct reconnaissance missions over the Baltic Sea just outside Soviet air space, which would penetrate close to Leningrad. Part of the idea of this dangerous mission, which is not believed to have been flown, was to provoke and evaluate Soviet air defenses in order to evaluate their radar effectiveness.

In 1962, DCI John McCone had promised President Kennedy that the A–12 was an alternative to reconnaissance satellites for spying on the Soviet Union. Nevertheless the A–12 is not known to have ever overflown the other superpower—and certainly not the heavily defended air space around Moscow and Leningrad. Indeed, as Corona satellites gradually improved, there was no need.

As 1965 faded into 1966, and 1966 into 1967, the A–12s continued to rest on the sidelines at Groom Lake. Lyndon Johnson, who had been so anxious to reveal the existence of the Oxcart during an election year, essentially kept it confined to its Area 51 barracks for three years. Indeed, there was even talk of putting the fleet into storage.

Finally, though, it was time for the Oxcart to fly the missions for which it had been created. In May 1967, at the urging of DCI Richard Helms, President Johnson directly authorized Operation Black Shield, the deployment of a detachment of A-12s to the Far East. The aircraft flew from Groom Lake, making much of the trip at Mach 2, directly to Kadena AB on Okinawa for operational missions.

By now, the war in Southeast Asia was escalating, and a need grew for reconnaissance flights over North Vietnam, an environment that was considered to be increasingly dangerous for U-2 operations. The first Black Shield mission, flown by Mele Vojvodich, took place on May 31, with the A-12 making a roughly 5,000-mile round trip to the demilitarized zone (DMZ) in three and a half hours, cruising at Mach 3–plus and 80,000 feet.

Through the end of 1967, three A-12s (Articles 127, 129, and 131) had flown twenty Black Shield missions over North Vietnam and two over Laos and Cambodia. During these flights, the A-12s were tracked by enemy radar on several occasions and fired on at least three times by a SAM. Because of the aircraft's speed and countermeasures, this shot fell far short.

One A-12 pilot is recalled to have quipped, paraphrasing Psalm 23, that "though I fly through the Valley of Death, I shall fear no evil for I am at 90,000 feet and climbing."

While the A-12 embodied aircraft technology is still impressive today, the photographic technology may seem like something from a distant, dark age. Many people—arguably most people—now carry in their pockets devices that can take pictures and instantly transmit them to virtually any place on earth. Once taken, the Black Shield photographs of 1967 were flown by courier to the headquarters of Eastman Kodak in Rochester, New York, for processing. Only then could they be delivered, by hand, to Langley, Virginia. Later, Black Shield film was processed at Yokota AB in Japan.

During the first quarter of 1968, four missions over North Vietnam were complemented by two over North Korea in the wake of the North Korean seizure of the surveillance ship USS *Pueblo* on January 23. Indeed, it was an A-12 flown by Jack Weeks that located the captured ship in Wonsan Harbor.

With plans already underway for CIA A-12 operations to be superseded by US Air Force SR-71 flights, the last of twenty-nine Black Shield missions was flown over North Korea on May 8, 1968. The long awaited, and long delayed, service career of the A-12 ended after less than one year.

Above: A CIA A-12 closes on a US Air Force KC-135 tanker over the Sierra Nevada. *Terry Panopalis collection*

Right: The A-12B Titanium Goose is shown parked near the Groom Lake tower. *Tony Landis collection*

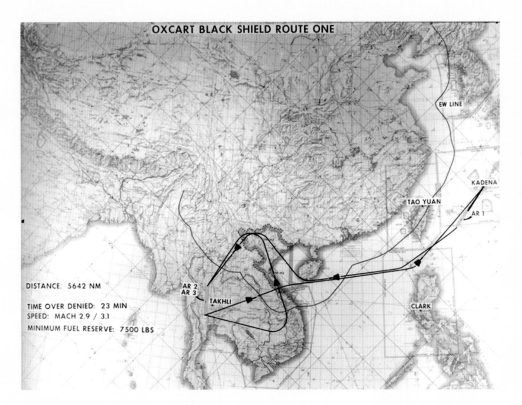

OXCART BLACK SHIELD ROUTE ONE

EW LINE

KADENA

TAO YUAN

AR 1

DISTANCE: 5642 NM

TIME OVER DENIED: 23 MIN
SPEED: MACH 2.9 / 3.1
MINIMUM FUEL RESERVE: 7500 LBS

AR 2
AR 3

TAKHLI

CLARK

This CIA map illustrates the scope of Black Shield operations over North Vietnam, with airfields available to A-12 Oxcarts marked. The red line indicates the extent of North Vietnamese and Chinese Early Warning radar. The yellow rectangles indicate aerial refueling tracks. *CIA*

That last mission was flown by Jack Layton in Article 131 (tail number 60-6937). Coincidentally, it was the same aircraft that had flown the first A-12 mission in May 1967.

The SR-71s began arriving at Kadena in March and flew their first operational mission on March 21. Operations of the two aircraft, and their respective CIA and US Air Force detachments, overlapped for about six weeks. DCI Richard Helms wanted to see the CIA A-12 missions continue, but his voice was drowned out by those in the Defense Department—especially Defense Secretary Robert McNamara—and the Bureau of the Budget who found the duplication of effort to be very cost inefficient.

An inflight view of an A-12 shows how it would have looked over North Vietnam. Faster than a speeding bullet, the Oxcart could outrun SAMs. *Terry Panopalis collection*

The three A-12s then based at Kadena were under orders to redeploy back to Groom Lake on June 8, but four days before, one of these was lost over the Pacific Ocean east of the Philippines while breaking in a replacement engine. No trace was ever found. The two combat survivors joined the other eight remaining never-deployed A-12s, and they were all transferred from Groom Lake to storage at the Skunk Works facility at Palmdale, California. This author recalls touring this boneyard in 1987 and being warned to take no pictures. This mothball fleet has since been parceled out to air museums around the United States.

As the A-12 era slipped farther into the past, not all of its secrets have been revealed. Most notable among those that were later revealed was its use as a launch vehicle for yet another, much blacker, black airplane.

CHAPTER 8
THE TAGBOARD FROM THE SORCERER'S DUNGEON

LYNDON JOHNSON LIFTED the shroud on the A–12 and then on the SR–71, but he never mentioned—and may never have known about—the most secret, and highest performing, member of their family.

Back in October 1962, at the height of the Cuban Missile Crisis, Kelly Johnson made a suggestion to the CIA that addressed their concern about pilots being lost during dangerous deep-penetration, overhead reconnaissance missions. He proposed the development of a Mach 3 reconnaissance aircraft that carried no pilot.

At the turn of the next century, during the formative years of the concept of unmanned combat air vehicles (UCAV), those who articulated the need for such vehicles that could fight remotely without pilots aboard characterized their missions as being "the dull, the dirty and the dangerous." As was so often the case, Kelly Johnson was ahead of his time.

He imagined an unmanned drone to be launched from the back of an A–12 that was already flying at high speed and high altitude. Because it would be launched from

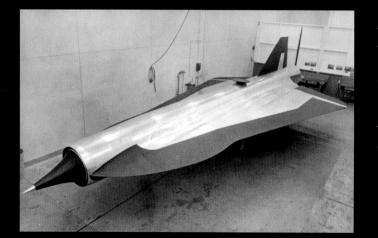

Unveiled at the Skunk Works in late 1963 to a select few CIA personnel who were on a "need-to-know" basis, the Q-12 mockup provided a first look at the D-21 configuration. *Lockheed*

another aircraft, this drone could use a ramjet engine, which is capable of higher speeds—above Mach 4—than a turbojet. As a trade-off for this speed, ramjets operated for a short duration, typically less than a half hour, so they had a very short range. However, if a ramjet-powered vehicle was carried by another aircraft for much of a long mission, the high speed would be available at the critical moment without worrying about the ramjet's short range.

The Skunk Works team was no stranger to ramjets, having used them between 1951 and 1960 in the X-7 series of experimental vehicles. The X-7 had achieved speeds up to Mach 4.3 and altitudes above 98,000 feet and later evolved into the AQM-60 Kingfisher target drone. Johnson turned to the Marquardt Corporation in Van Nuys, just a few miles west of Burbank, which had built the XRJ43-MA series ramjet engines for the X-7 test program. For the new drone, Marquardt would build the RJ43-MA-20S4 variant, which eventually proved itself to have a duration of around 90 minutes.

In true Skunk Works form, a full-scale mock-up was built in just six weeks and sent to Groom Lake in December 1962 for RCS tests. However, the CIA contract was not issued for another three months. General Leo Geary, who had long been a liaison between the CIA and US Air Force on black airplane projects, became the customer's point man for the program.

"With General Geary acting as our champion, the [CIA] decided to climb on board the drone project," Ben Rich explains. "On March 20, 1963, we were awarded a letter contract from the CIA, which would share funding and operational responsibilities with the Air Force."

The CIA assigned the code name Tagboard to a project that was undertaken with, if anything, even greater secrecy than that which had shrouded the Oxcart program.

"Tagboard now became the most classified project at the Skunk Works, even more secret than the Blackbird airplanes being assembled," Rich continues. "Kelly decided to

A D-21 Tagboard drone is attached to the M-21 Mothership, the last aircraft in the A-12 lineage. First flown in May 1965, this M-21 successfully launched three D-21s before colliding with a drone on July 30, 1966. Tagboard missions were launched from Groom Lake. *Tony Landis collection*

wall off a section of the huge assembly building housing Blackbird, which already was as guarded as Fort Knox, to accommodate the new drone project. To get inside that walled-off section required special access passes and the shop workers immediately dubbed it Berlin Wall West."

As a designator, Q-12 ("Q" being the standard military designation for drones) was initially assigned to this project, which was seen as a derivative of the A-12 program. However, during 1963, the numerals were transposed to "21" to avoid confusion with the A-12 series, the prefix was changed from "Q" to "D," and the vehicle became known as the D-21.

Some sources state that the "D" stood for "drone," while others say that it stood for "daughter" because the A-12s that were being specifically built to launch the new vehicle were designated with an "M" for "mother" or "mothership." The last two of the fifteen A-12 aircraft, Articles 134 and 135 (tail numbers 60-6940 and 60-6941), were built in 1963 and 1964 and officially designated M-21s. Like the YF-12A and SR-71, they had two cockpits: one for the pilot and one for the launch system operator.

The D-21 drone itself was 42 feet 10 inches long, with a wing span of 19.25 feet, and had a gross weight of 11,000 fully fueled at launch. Like the A-12 family, it was built of titanium. To save weight, it was unencumbered with landing gear, which was unnecessary. It was air-launched and designed for a one-time use, with only the nose section, containing the cameras and film pack, being recoverable. A total of thirty-eight D-21s would be built.

The US Air Force, meanwhile, kept an eye on this CIA program but shied away from participating, despite Kelly Johnson's efforts to sell additional D-21s to SAC. Part of his reasoning was that unit costs would be driven down with an increase in the number of units. He proposed that SAC would not have to acquire M-21s. They could use the Boeing B-52H Stratofortress to launch the drones.

Below: Before completion, the first D-21 was experimentally mated to the first M-21 at Burbank to make sure that everything was going to fit. The M-21 carried US Air Force markings as a cover. *Lockheed*

Above: A Marquardt XRJ43-MA-20S4 ramjet is installed in the first D-21. The Tagboard drones were marked only with their article number. *Lockheed*

Above: A D-21 is mated to an M-21 at Groom Lake. For captive test flights, the Tagboard drone was fitted with an aerodynamic tail cap. *Tony Landis collection*

Left: Ready for flight, the second M-21, with its D-21, waits on the Groom Lake runway. *Tony Landis collection*

After spending the better part of 1963 finalizing the aerodynamics and developing the camera system, the first mating of a D–21 to an M–21 took place at Plant B-6 in Burbank in June 1964. These aircraft were delivered to Groom Lake two months later. In his official log, Kelly Johnson notes that the mated pair "went supersonic on the first flight" on December 22.

A Tagboard D-21 flies under the wing of a B-52H. After the loss of the only operational M-21 Mothership in 1966, the Stratofortress was used as a "step-Mothership." *Tony Landis collection*

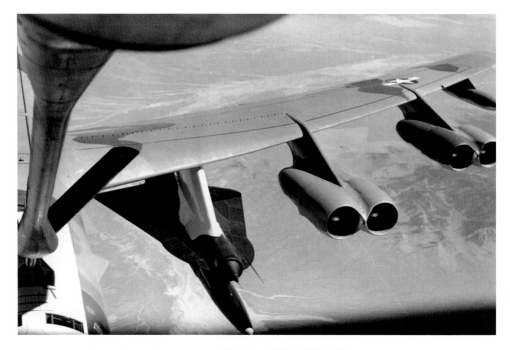

Left: The M-21 Mothership and a D-21 "daughter ship" are shown in the sky over the Nellis Range. The pair was first flown in December 1964 and at Mach 2 early in 1965. *Lockheed*

Johnson had high hopes for a first D-21 launch in early 1965, perhaps even on his February 27 birthday. However, this did not happen until 1966. Throughout 1965, the Tagboard test program was consumed with flight testing a paired vehicle that was substantially different aerodynamically than the sum of its parts. Typical test flights took off from Groom Lake and passed over the Pacific Ocean by way of Naval Air Station Point Mugu on the California coast, where the US Navy has conducted its missile tests since the 1940s.

Though the M-21/D-21 pair achieved speeds in excess of Mach 2, there were problems such as transonic acceleration and the still untested issue of separating a 5.5-ton vehicle off the back of an M-21 at such speeds. In his log, Kelly Johnson admitted that the separation, with the engines of both D-21 and M-21 accelerating, was "the most dangerous maneuver we have ever been involved in [with] any airplane I have ever worked on."

"The technical problems were formidable, especially the attempt to launch a piggybacked drone from a mothership launch platform flying at three times the speed of sound," Ben Rich recalled. "The drone would be sitting toward the top rear of the fuselage on a pylon. Expecting a drone to launch through the mothership's Mach 3 shock wave presented a monumental engineering challenge. And Kelly insisted that we launch at full power. It took us nearly six months to work out some of the shock wave and engine problems with models in the wind tunnel, while other problems concerned with perfecting the guidance system, the cameras, the self-destruction system, the parachute deployment system, all loomed before us like monsters let out of some evil sorcerer's dungeon."

The first separation, and therefore the first D-21 launch, finally took place on March 5, 1966, with Lockheed test pilots Bill Park and Keith Beswick in the M-21's tandem cockpits. This merely tested the launch procedure, but additional test flights were aimed at flight testing the D-21 vehicle. The April 27 flight found the D-21 accelerating to Mach

A D-21 is shown rotated sideways during the conversion to D-21B standard for Senior Bowl operations. *Lockheed*

The first D-21 is shown at the Skunk Works being modified as a D-21B. It was later lost in an accidental drop. *Lockheed*

3.3, reaching 90,000 feet and following a programmed course for 1,380 miles. On June 16, another D-21 flew a preprogrammed, 1,840-mile course with multiple programmed turns.

The turning point for Tagboard came on July 30. It might have been a confirmation of a promising career, but it marked the end of a program that had barely begun. Bill Park was piloting the number two M-21 (60-6941), which had been used for all the successful launches to date, with Ray Torick in the back seat. As with the previous tests, the other M-21 was flying as a chase plane for the operation.

Shortly after it was launched at Mach 3.3 over the Pacific off Point Mugu, in that most dangerous moment of the "most dangerous maneuver," the D-21 struck the mothership and the two aircraft cartwheeled out of the sky. Both crewmen successfully ejected, but Torick was injured and drowned upon landing in the ocean.

There would never be another M-21/D-21 launch, but Kelly Johnson had already proposed the switch to a B-52H as mothership. The B-52H was much slower than an A-12 or M-21, so the drone could not be launched at supersonic speeds, but there was the safety advantage of being able to *drop* the D-21. The D-21 launched from an M-21 by accelerating off its back, so gravity was not on the side of this maneuver.

The slower speed of the B-52H, meanwhile, meant that the D-21 was not immediately supersonic at launch, so the ramjet was augmented by a rocket that could accelerate the vehicle to Mach 3.3. The thirty-two remaining D-21s were modified at Burbank for drop launches and redesignated as D-21Bs. Both the drones and motherships were extensively modified with a navigation and communications interface so that a drone could be controlled and directed by the mothership.

Under the operational name Senior Bowl, the US Air Force created the 4200th Support Squadron at Beale AFB in California and promptly relocated it to Groom Lake. Beale had previously been home to the 4200th Strategic Reconnaissance Wing (SRW), which had been established in January 1965 to receive US Air Force SR-71s and which was redesignated as the 9th SRW ten months later. The 4200th Support Squadron was responsible for the D-21 program at Groom Lake while the 9th SRW remained at Beale. Two B-52Hs were assigned to the 4200th Support Squadron and sent to the Skunk Works facility at Palmdale for modifications.

One would have expected that with gravity on its side, the B-52H/D-21B launch scenario would have been more successful than Tagboard launches had been, but this was not the case. Whereas the M-21/D-21 pairing had succeeded in its first three launch attempts, a number of aborts and launch failures preceded the first successful Senior Bowl B-52H/D-21B launch. This successful launch took place on June 16, 1968, two years after the M-21 flights ended and a year after the last A-12 was retired.

The program moved slowly, but after three more successful tests during 1969, the 4200th Support Squadron deployed overseas to Andersen AFB on Guam to begin operation reconnaissance flights. The stated target was the Chinese nuclear weapons test facility at Lop Nur deep inside the Xinjiang Uygar region in northwestern China. The Chinese

The seventh D-21 is shown being used as a testbed to develop the B-52H launch pylon. Article 507 was test flown on November 6, 1967. *Lockheed*

had detonated their first nuclear weapon there in 1964 and their first hydrogen bomb in 1967.

The first operational Senior Bowl mission finally took place on November 9, 1969, seventeen months after the first successful flight. The D–21B succeeded in overflying Lop Nur without detection, but it failed to execute a programmed 180-degree turn and continued flying on until it ran out of fuel over the Soviet Union

and crashed. The wreckage was recovered and used to design a reverse-engineered copy that was never built.

The second mission on February 21, 1970, succeeded in making the turn, but a parachute failure caused the film package to be lost in the Pacific. The final two known D–21 missions occurred on March 4 and March 20, 1971. Both were failures. In the first instance, the flight went well but the film capsule was hit by a recovery ship and lost. The other mission ended when the D–21B went down over China. The wreckage went on display at the China Aviation Museum near Beijing in 2010.

In July 1971, four months after the latter mission, the CIA formally terminated Tagboard. The surviving D–21Bs were secretly slipped out of Groom Lake and taken to an obscure corner of the US Air Force boneyard at the Military Aircraft Storage and Disposition Center (MASDC)—later called the Aerospace Maintenance and Regeneration Center (AMARC) and now the Aerospace Maintenance and Regeneration Group (AMARG)—at Davis-Monthan AFB near Tucson, Arizona.

Had they not been accidentally discovered at the boneyard by a civilian journalist in 1977, they would never have been acknowledged publicly and this black aircraft would be one more whose existence was shrouded in perpetual darkness.

CHAPTER 9

THE BLACKBIRD

DEVELOPED UNDER THE code name Senior Crown, the SR-71 Blackbird became the ultimate Archangel, the capstone in the lineage that began with the first A-12. The SR-71 has the distinction of having served for more than three decades, while the A-12 was in combat for barely a year. No other aircraft has ever had the distinction of being the fastest operational aircraft in the world from the day it entered service until the day it was retired three decades later. No other aircraft has ever set a world speed record on its retirement flight.

In 1983, in a flightline conversation at Beale AFB, an SR-71 pilot told this author that the Blackbird represented "high nineties technology that we were lucky to have in the sixties." Today the nineties have come and gone, but there has yet to be anything else quite like the extraordinary Blackbird.

"The Blackbird was a wild stallion of an airplane," Ben Rich, the program manager, recalled in his memoirs. "Everything about it was daunting and hard to tame— building it, flying it, selling it. It was an airplane so advanced and awesome that it easily intimidated anyone who dared to come close. Those cleared to see the airplane roar into the sky would remember it as an experience both exhilarating and terrifying as the world shook loose . . . with the roar of an oncoming tornado and the ground shaking under [one's] feet like an eight-point earthquake, as the engines spouted blinding diamond-shaped shock waves."

One of those "cleared to see" the SR-71 was CIA Director Richard Helms.

"I was so shaken, that I invented my own name for the Blackbird," Helms later told Ben Rich about watching a nighttime launch at Groom Lake. "I called it the Hammers of Hell."

Five feet longer but largely similar to the single-seat A-12, the tandem seat SR-71 evolved out of Kelly Johnson's suggestion that the US Air Force should consider a reconnaissance aircraft like the CIA's Archangel. While the A-12 and YF-12A aircraft were originally delivered mainly in a natural metal finish, SR-71s were coated entirely in a dark blue-black paint, earning them the Blackbird name.

The first SR-71A (tail number 61-7950) made its debut flight at Palmdale, California, near Edwards AFB, on December 22, 1964. Lockheed test pilot Bob Gilliland, a veteran

A 9th Strategic Reconnaissance Wing SR-71 Blackbird. *USAF, Tech Sergeant Michael Haggerty*

Three SR-71 Blackbirds take shape on the factory floor at Lockheed's top secret Plant B-6 in the late 1960s. *Lockheed*

Left: This rear view provides a good look at the titanium inner structure within the wings of the SR-71 Blackbird. *Lockheed*

of the A–12 program, was at the controls. The second and third Blackbirds made their first flights during March 1965.

A total of thirty-one Blackbirds rolled out of final assembly at Palmdale between August 1964 and May 1967. These included twenty-nine SR–71As and two SR–71Bs, the latter designed as trainers with an elevated rear seat in a fashion similar to that of the A–12B Titanium Goose. In the SR–71A, unlike the A–12B and the SR–71B, the rear seat, accommodating the reconnaissance systems officer (RSO), was not elevated.

In addition to the A and B variants, a thirty-second Blackbird was designated as SR–71C, which was completed in 1969 using the salvaged rear section of a YF–12A.

In January 1965, as a home for the incoming Blackbirds, the US Air Force activated the 4200th SRW at Beale AFB as a component of SAC. The subsidiary 4200th Support Squadron (later 4200th Test Wing) was the umbrella organization for the D–21 program at Groom Lake. In October 1965, the 4200th SRW was redesignated as the 9th SRW, assuming the lineage of the 9th Bombardment Group, which dated back to before World War II. This wing was comprised of two strategic reconnaissance squadrons (SRS), the 1st

The large edifice in the center of the Building 309-310 complex appears to be the frame of the crate used to ship large fuselage components from Burbank to Groom Lake. *Lockheed*

Portions of three SR-71 Blackbirds are shown under construction within the Building 309-310 complex at Plant B-6 in Burbank. *Lockheed*

SRS and 99th SRS. In July 1976, in a strategic reconnaissance consolidation, the U-2s of the 100th SRW were reassigned to the 9th SRW.

Aerial refueling support was initially provided to the 9th SRW by KC-135Q tankers operated by the 9th and 903rd Aerial Refueling Squadrons (ARS) of the 456th Bombardment Wing. After 1975, the squadrons were reassigned directly to the 9th SRW.

The first Blackbird to arrive at Beale AFB was an SR-71B trainer that came in on January 7, 1966. The first operational SR-71A reconnaissance bird arrived on April 4. The first overseas deployment came two years later, by which time all of the SR-71As and SR-71Bs had been delivered.

Even before the aircraft had much of a chance to prove themselves, the Nixon administration counterintuitively decided that there should not be more Blackbirds—ever. They went so far as to demand that Lockheed literally *break the mold*. Aside from the single SR-71C hybrid, no more Blackbirds were built.

"One of the most depressing moments in the history of the Skunk Works occurred on February 5, 1970, when we received a telegram from the Pentagon ordering us to destroy all the tooling for the Blackbird," Ben Rich recalls sadly. "All the molds, jigs, and forty thousand detail tools were cut up for scrap and sold off at seven cents a pound. Not only didn't the government want to pay storage costs on the tooling, but it wanted to ensure that the Blackbird never would be built again. I thought at the time that this cost-cutting decision would be deeply regretted over the years by those responsible for the national security. That decision stopped production on the whole series of Mach 3 aircraft for the remainder of [the twentieth] century. It was just plain dumb."

Indeed, the fascinating career of the Blackbird had barely begun.

Beginning on March 8, 1968, the 9th SRW formed a detachment of Blackbirds at Kadena AB on Okinawa, where they operated alongside the CIA A-12 detachment until May 8. Nicknamed "Habu" after a pit viper indigenous to Okinawa, the SR-71s would remain at Kadena for more than two decades until early 1990. During most of this time, they were known as Detachment 1, although they were originally called OL-8 (for Operating Location 8, numbered in sequence with previous SAC U-2 detachments).

Another SR-71 nickname that came into use was the term "Sled," which was widely used by Blackbird pilots, who referred to themselves as "Sled Drivers."

The Kadena detachment's first mission on March 21, 1968, was followed by 167 more through the end of the year. The numerous wartime missions through the next few years included key battlefield surveillance missions, including those that helped planners assess air support for major battles, including the siege of Khe Sanh.

Other missions were conducted over North Korea and the periphery of both Chinese and Soviet air space—the latter including surveillance of the Soviet naval facilities around Vladivostok. Detachment 1 also conducted long-range missions over the Middle East during the Iran-Iraq War of the 1980s.

Detachment 4 of the 9th SRW was established at RAF Mildenhall in the United Kingdom in 1976, hosting short duration SR-71 and U-2 deployments until 1984, after which it became a permanent fixture through 1990. Missions included routine surveillance of East Germany, Poland, the Baltic Sea, and Soviet bases on the Kola Peninsula. In April 1986, Detachment 4 Blackbirds conducted pre- and poststrike reconnaissance of Libyan targets that were attacked during Operation El Dorado Canyon.

The 9th SRW also operated SR-71 missions directly from the United States. In 1973, they conducted overflights of the Middle East during the Yom Kippur War, staging from Beale by way of Griffis AFB in New York and Seymour Johnson AFB in North Carolina.

Under operational code names including Giant Plate and Clipper, the 9th SRW conducted routine overflights of Cuba through the 1970s. Unlike the more vulnerable U-2s, the fast, high-flying SR-71s were essentially impervious to any form of air defenses that could be brought to bear over Cuba.

During the 1970s, the US Air Force authorized the SR-71 to come out of the shadows long enough to give the world a sense of its capabilities. On September 1, 1974, Major James Sullivan and Major Noel Widdifield set the speed over a recognized course record while flying 3,508 miles from New York to London in just under two hours at an average speed of 1,435.6 mph.

In this view, the internal structure of the chines, which run along the SR-71 forward fuselage, are visible. *Lockheed*

Above: This Blackbird, the seventh off the line, was the first of two SR-71B two-seat trainers. It is seen here in this photo from 1982 on its 1,000th sortie. *USAF, Staff Sergeant Bill Thompson*

Left: Kelly Johnson chats with Morley Safer during an interview about the Blackbird, which aired on the *60 Minutes* television program in October 1982. *Lockheed*

On July 27 and 28, 1976, three SR-71s were used to set three separate absolute world records. Captain Robert Helt and Major Larry Elliott set the record for absolute altitude in horizontal flight (by an aircraft taking off under its own power) of 85,069 feet. Major Adolphus Bledsoe and Major John Fuller set an absolute closed course speed record of 2,092.3 mph. Finally, Captain Eldon Joersz and Major George Morgan set an absolute straight course speed record of 2,193.2 mph that still stood in the second decade of the twenty-first century.

The Blackbird's full potential of speeds in excess of Mach 3.3 and operations above 100,000 feet has been repeatedly rumored but never made part of any official record.

The actual top speed of the SR-71 is still classified. Some people say that it was far beyond Mach 3.3. Others have said that it was never reached, that the Blackbird never was accelerated to its full potential maximum speed. An SR-71 pilot once told this author that if any other aircraft ever took away the Sled's absolute speed record, one of

An SR-71 Blackbird assigned to Detachment 4, 9th SRW, prepares for takeoff from RAF Mildenhall in 1983. *USAF, Tech Sergeant Jose Lopez*

An SR-71 takes off from a mist-shrouded runway at RAF Mildenhall in England for an undisclosed mission over Europe or the Baltic. *USAF, Tech Sergeant Jose Lopez*

the 9th SRW pilots would just go up the next day and "step down a little harder on the accelerator."

The record still stands.

In another conversation, this author was speaking with a former ground radar operator who tracked an aircraft, not a missile, flying at *Mach 6*, and he nearly panicked. If there was ever a case of a truly unidentified UFO, this was it. The man reported this bogey to the officer in charge, who glanced at the scope and assured him, "Don't worry, it's one of ours."

In his book, *Sled Driver*, SR-71 pilot Brian Shul recalled a radio exchange that occurred as he was over Southern California at 68,000 feet. Monitoring various radio transmissions from other aircraft, he heard a Cessna ask for a readout of its groundspeed.

"Ninety knots," replied Los Angeles Center.

A Twin Beech asked for the same and was given a faster speed of 120 knots.

At that moment a cocky Navy F/A-18 pilot came on.

"Center, Dusty 52 requests groundspeed readout."

The response came, "525 knots on the ground, Dusty."

Unable to resist, Shul and his RSO clicked their radios simultaneously.

"It was at that precise moment I realized Walt and I had become a real crew," Shul recalls. "We were both thinking in unison.

"Center, Aspen 20," Shul said, addressing Los Angeles Center. "You got a ground speed readout for us?"

"Aspen," the controller replied after a long pause. "I show 1,742 knots."

Shul notes that "no further inquiries were heard on that frequency."

Though the SR-71 was probably never seriously threatened by enemy countermeasures, its ultimate undoing was, ironically, another Lockheed product, which was *not* an airplane.

As Lockheed's Skunk Works was building spyplanes for the CIA, Lockheed Space Systems was developing spy *satellites* for the National Reconnaissance Office (NRO). During the Cold War, if there was anything blacker in the metaphorical sense than the

CIA and the black jets of Area 51, it was the NRO and its satellites. These were operated under the cover name "Discoverer," but were known in the black world as "Keyhole" after their Itek high-definition cameras. Indeed the NRO itself, and the work it was doing in the 1960s and 1970s, was not declassified until the 1990s. Information about the work it is doing today is not something for which one should hold one's breath.

The NRO was formed in suburban Washington, DC, in 1961 specifically to centralize work being done by the CIA and DOD to develop reconnaissance satellites. The NRO was separate from the CIA, although there would be extensive interaction, and many former CIA and black world spyplane hands, such as Ozzie Ritland and Richard Bissell, played a role in NRO's early days.

Lockheed Space Systems and the Lockheed Missile Division, which were later combined to form the Lockheed Missiles & Space Company (LMSC), were created in Southern California but moved north in the late 1950s to what later became Silicon Valley, finally settling in Sunnyvale. It was responsible for the Polaris, Poseidon, and Trident submarine-launched missiles, as well the NRO spy satellites.

During the Cold War, Blackbirds of the 9th SRW deployed around the world from their stateside base at Beale AFB. *USAF, Tech Sergeant Michael Haggerty*

The Discoverer/Keyhole series included the KH-1 through KH-3 satellites, which were part of a program code named Corona. Also coming under the NRO mandate were the KH-4 Mural, KH-5 Argon, and KH-6 Lanyard. Operational through the 1960s and into the 1970s, the early Keyholes were "film-return" systems in which photographic film was dropped back into the atmosphere from outer space, retrieved by specially modified aircraft, and processed. Through 1972, the KH-1 through KH-6 spacecraft exposed 2.1 million feet of film and took 800,000 pictures.

In many ways, the early Keyholes were operationally inferior to the SR-71 and its fellow Archangels. While the resolution of the cameras was the best that money could buy, the satellites orbited 75 to 100 miles above their subjects, while Blackbirds flew less than 18 miles above. Aircraft could also be sent over a specific target at a specific time, while satellites were confined to specific orbits. Finally, the process of retrieving the film capsules was complicated, difficult, and not always certain, despite techniques having been honed to a fine art by those doing the retrieving.

All this began to change late in 1976, as the NRO deployed the first of its KH-11 satellites, which now used electro-optical digital imaging. As the KH-11 satellites matured, and as at least a half dozen were launched during the 1980s, photoreconnaissance changed completely. No longer did film have to be retrieved, and no longer did decision makers have to wait days to see their coveted secret pictures. They could now see them instantaneously.

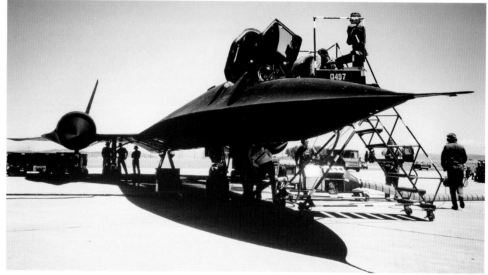

Top and right: Cranking up an SR-71A Blackbird for launch from Beale AFB in California in June 1983. *Bill Yenne*

Despite the retrofitting of digital systems and communications links aboard the SR-71s, which allowed them to deliver imagery in near "real time," the US Air Force itself recommended the retirement of the Blackbirds.

"General Larry Welch, the Air Force chief of staff, staged a one-man campaign on Capitol Hill to kill the program entirely," Ben Rich wrote in his memoirs. "General Welch thought sophisticated spy satellites made the SR-71 a disposable luxury. Welch had headed the Strategic Air Command and was partial to its priorities. He wanted to use SR-71 refurbishment funding for development of the B-2 bomber. He was quoted by columnist Rowland Evans as saying, 'The Blackbird can't fire a gun and doesn't carry a bomb, and I don't want it.' Then the general went on the Hill and claimed to certain powerful committee chairmen that he could operate a wing of fifteen to twenty [F-15E] fighter-bombers with what it cost him to fly a single SR-71. That claim was bogus. So were claims by SAC generals that the SR-71 cost $400 million annually to run. The actual cost was about $260 million."

Both Welch and SAC commander General John Chain testified before Congress that the SR-71 should go, and so it did.

As Rich so aptly reflected, "a general would always prefer commanding a large fleet of conventional fighters or bombers that provides high visibility and glory. By contrast, buying into Blackbird would mean deep secrecy, small numbers, and no limelight."

Blackbird operations, except training flights, were officially terminated in November 1989, having been eliminated from the FY1990 Defense Department budget.

On March 6, 1990, one Blackbird famously set a series of world speed records on its "retirement flight." The SR-71 with tail number 64-17972 was flown from California to the Smithsonian National Air & Space Museum (NASM) Udvar-Hazy Center at Dulles Airport, where it would eventually go on display. In the process, it set the official National Aeronautic Association coast-to-coast speed record of 2,086 miles in one hour and seven minutes, averaging 2,124.5 mph. It made the 311-mile St. Louis-to-Cincinnati leg in less than nine minutes, averaging 2,176.08 mph.

Within a few months of this much-publicized flight, Saddam Hussein's Iraqi

Left: Wearing his high altitude flight suit, an SR-71A pilot prepares for a 1983 mission. *Bill Yenne*

Below: An SR-71 Blackbird touches down after a long mission. *NASA*

army had occupied Kuwait and the United States was involved in the Desert Shield buildup that culminated in Operation Desert Storm in January and February 1991. During that conflict, many operational commanders, including General Norman Schwarzkopf, lamented the absence of expedited reconnaissance that the SR-71 might have contributed.

Mounting concerns about the situations in world trouble spots from the Middle East to North Korea led Congress to reconsider the reactivation of the SR-71. In 1993, Admiral Richard Macke, director of the joint staff for the Joint Chiefs of Staff, told Congress that "from the operator's perspective, what I need is something that will not give me just a spot in time but will give me a track of what is happening. When we are trying to find out if the Serbs are taking arms, moving tanks or artillery into Bosnia,

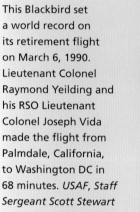

we can get a picture of them stacked up on the Serbian side of the bridge. We do not know whether they then went on to move across that bridge. We need the [reconnaissance information] that a tactical, an SR-71, a U-2, or an unmanned vehicle of some sort, will give us, in addition to, not in replacement of, the ability of the satellites to go around and check not only that spot but a lot of other spots around the world for us. It is the integration of strategic and tactical."

In its FY1994 appropriations, Congress authorized a reinstatement of funding to

This Blackbird set a world record on its retirement flight on March 6, 1990. Lieutenant Colonel Raymond Yeilding and his RSO Lieutenant Colonel Joseph Vida made the flight from Palmdale, California, to Washington DC in 68 minutes. *USAF, Staff Sergeant Scott Stewart*

Note the Skunk Works logo on the Linear Aerospike SR Experiment (LASRE) pod on the back of a NASA SR-71 in this photo from February 15, 1996. *NASA, Tony Landis*

permit a revival of part of the SR-71 fleet. By that time, many of the twenty surviving SR-71s were being prepped for museum displays, but at least a half dozen were in storage at Palmdale or flying research missions with NASA.

The US Air Force moved too slowly on the path to SR-71 reactivation, and in October 1997, using a line-item veto, President Bill Clinton deleted the funding. The Blackbird was permanently grounded by the US Air Force in 1998, leaving just two at NASA's Dryden Flight Research Center at Edwards AFB.

One of the last NASA missions for the SR-71 was the Linear Aerospike SR-71 Experiment (LASRE) series conducted in 1997 and 1998. The object was to study aerodynamic performance of lifting bodies combined with aerospike engines such as would have been used in the Lockheed Martin Skunk Works X-33, the demonstrator for the conceptual VentureStar single-stage-to-orbit reusable spaceplane. The latter program was abandoned by NASA in 2001 but pursued by Lockheed Martin thereafter.

In signing off any discussion of the Blackbird's demise, Americans are left with the words that Senator John Glenn spoke on the floor of the US Senate on the day after the 1990 "retirement flight."

Said the former astronaut, "The termination of the SR-71 was a grave mistake and could place our nation at a serious disadvantage in the event of a future crisis. Yesterday's historic transcontinental flight was a sad memorial to our short-sighted policy in strategic aerial reconnaissance."

The NASA SR-71B Blackbird in flight over the Sierra Nevada in 1994. *NASA, Judson Brohmer*

CHAPTER 10
MiGS OVER TONOPAH

IF GROOM LAKE, Area 51, Paradise Ranch, or whatever it's really called is in the middle of nowhere, then Tonopah is at a distant, mainly forgotten corner of nowhere.

The folklore of the Old West is peppered with stories of prospectors who found their gold strikes while searching for lost burros. In 1900, this actually happened on the site of Tonopah, and that's how the town got its start. By 1920, though, the mines played out, and so too—almost—did Tonopah. Were it not for a World War II–era USAAF fighter training base nearby, and the fact that Tonopah is one of the few places to stop on a lonely, 500-mile stretch of US Highway 95 north of Las Vegas, Tonopah would have been a ghost town by 1940. If it had not been for this airfield, which later fell within the borders of the vast NTTR, Tonopah might very well be a ghost town now.

The story of how an airport 30 miles southeast of an old almost–ghost town came to be the home to leading edge warplanes from the Soviet Union represents one longest kept secrets in the history of the US Air Force—and in the history of Area 51 and the NTTR.

Even today, long after this hidden program was officially acknowledged, official information remains sparse. It will likely remain that way.

The story begins during the Vietnam War, at a time when US Air Force and US Navy pilots found themselves in the first aerial combat that the services had experienced since the Korean War. Only a dozen years had passed, but in those dozen years, there had been huge changes in air combat doctrine and in fighter aircraft. It was supposed that future air combat would be conducted at great speed, great distance, and through the use of air-to-air missiles. It was believed that air combat maneuvering (ACM), i.e., dogfighting, had been rendered obsolete by air-to-air missiles that made beyond visual range (BVR) missile engagements possible.

Combat over North Vietnam proved the error of that assumption. Fighter pilots who had never been trained for the type of combat experienced by their forebears in World War II and Korea had a rude awakening. Pilots flying fast fight aircraft unarmed with guns found themselves in old-fashioned dogfights with nimble, highly maneuverable, Soviet-built fighters with guns that were ideally suited for dogfight tactics—where missiles were a far less desirable weapon.

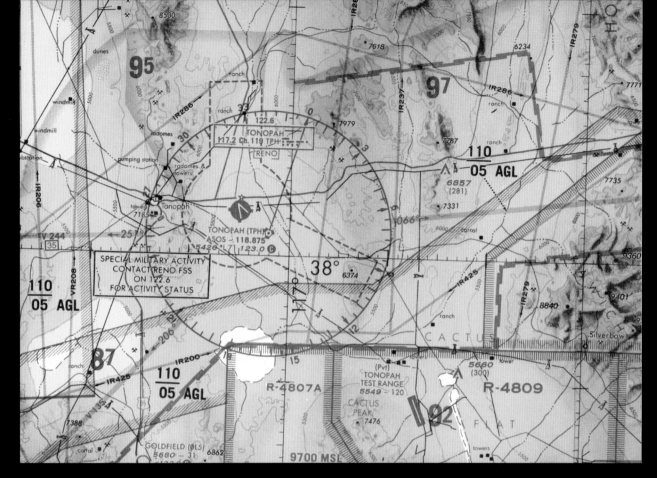

The US Navy and US Air Force TAC scrambled to make changes. These included arming existing fighters with guns, specifying guns in future fighters, and revamping training. The latter included the US Air Force Red Baron program, which began studying enemy ACM tactics in 1966, although the US Navy was the first to formally undertake a training program. In 1969, they created their Navy Fighter Weapons School at NAS Miramar in California and initiated their Topgun (aka Top Gun) realistic fighter training exercises. It was not until 1975 that the Air Force's own analogous exercise, called Red Flag, became a permanent part of the curriculum at the US Air Force Fighter Weapons School (FWS) and Tactical Fighter Weapons Center (TFWC) at Nellis AFB. In 1976, similar exercises were begun at Clark AB in the Philippines under the name Cope Thunder.

The theory behind the exercises was the deduction, drawn from combat experience, that a fighter pilot was most likely to be defeated during his first ten air-to-air battles. If he survived the learning curve of those ten, he had an excellent chance to survive the rest of his combat career. The idea behind realistic fighter training was to give pilots their first ten "battles" in a controlled environment. Units called aggressor squadrons were formed, in which experienced pilots were trained to use enemy tactics. For the air force, the aggressors were set up under the TFWC's 57th Fighter Weapons Wing as the 64th Fighter Weapons Squadron.

Groom Lake runways appear on no unclassified government-issued maps, but the Tonopah Test Range runway does. Tonopah Airport is just east of town. The Tonopah Test Range field to the southeast is within the restricted airspace of Range R-4809. *FAA*

The aggressors would then fly against other pilots to give them a feel for flying against Soviet and Soviet-trained air forces. The aggressors flew aircraft, such as the Northrop F-5, which simulated the performance characteristics of Soviet aircraft, especially the ubiquitous MiG fighters.

The latter begs the question, why were real MiGs not used? The stock answer was that real MiGs were impossible to come by. The *real* answer was that American pilots *were* flying against real MiGs, and this fact was a secret for nearly two decades.

MiGs were the product of the Mikoyan-Gurevich Design Bureau, abbreviated as "MiG," that was formed before World War II under the direction of engineers Artem Mikoyan and Mikhail Gurevich. After several forgettable piston engine aircraft, Mikoyan and Gurevich achieved a place in aviation history with their MiG-15, which first flew in 1948. The two designers solidified their place in history with their MiG-17 and the more advanced, supersonic, MiG-19, which first flew in 1950 and 1953, respectively. The MiG-21, first flown in 1955, was produced in larger numbers (nearly 11,000) than any jet in history and was the backbone of many air forces—including North Vietnam's—for decades.

The MiG-15 came as a rude surprise during the Korean War to US Air Force pilots who found it a worthy opponent to the F-86 Sabre. As the very term "MiG" represented the state of the art in Soviet warplane technology, the United States was naturally anxious to get hold of one, and a $100,000 bounty was offered to potential defectors. It was not until September 1953, after the war ended, that a North Korean pilot, No Kum-Sok, defected with a MiG-15, and it was many years before the Americans got their hands on more advanced types. The MiG-15 made headlines. The later acquisitions came under the shroud of secrecy.

The surreptitious MiG collection began in 1968, even as American pilots were learning the hard lessons of ACM over North Vietnam. The first aircraft in the Nellis flock came from Israel, where Iraqi pilot Captain Monir Radfa defected with a MiG-21 in 1966. This aircraft was evaluated by the Foreign Technology Division (FTD) of the Air Force Systems Command (AFSC) under the code name Have Doughnut. The first word was used by the AFSC to identify its test programs, while the latter is said to be a reference to the circular shape of the MiG-21 gunsight.

During 1968, the Air Force acquired Syrian MiG-17s from the Israelis, as well as one from a Cambodian pilot who flew it to South Vietnam. These were evaluated by the FTD under the code names Have Ferry, Have Drill, and Have Privilege. Some of these aircraft were also taken to Groom Lake for evaluation by both Navy and Air Force combat pilots through the TFWC. As time went on, the MiG collection was being secretly expanded, including the addition of four MiG-21s from Indonesia—in trade for newer American F-5s—in 1973. By this time, the Area 51 MiGs were being test flown under the umbrella of a program called Have Idea. Fighter pilots, including Red Flag participants, were routinely taken to Groom Lake to see the MiGs, but it was not until 1975 that the US Air Force began organizing its MiGs into what amounted to an aggressor squadron that flew real, not simulated, Soviet bloc equipment.

The entrance to the Tonopah Test Range, circa 1980s. The Sandia National Laboratories, responsible for many of the activities occurring farther south in the Nevada Test Range, were an obvious choice for the cover story. *Lockheed*

Looking eastward along North Main Street shows Tonopah as it was at the time that the Aquatone U-2s had been glistening specks high overhead. *Author's collection*

Looking up Brougher Avenue at North Main Street. The Tonopah Club closed around the time that Red Flag brought life to the skies above, compelling thirsty flyboys from the TTR to drift across the street to the bar at the Mizpah Hotel. *Author's collection*

Under the Constant Peg program, the US Air Force began training full-time MiG pilots—from both its own ranks and from the US Navy and US Marine Corps—as well as translating manuals and setting up a maintenance infrastructure. The MiG collection was transformed from an FTD curiosity to an operational unit. Formed in April 1977, the MiG unit became the 4477th Test & Evaluation Flight (later Test & Evaluation Squadron), known as the "Red Eagles."

The MiG-17s and MiG-21s of the 4477th made their first and very secret appearance at a Red Flag exercise in the summer of 1977. The pilots chosen to fly against the MiGs were sworn to secrecy. Constant Peg was one of the most classified programs in the service. Even many senior air force commanders were not aware of it.

As more MiGs were added to the program—including a baker's dozen MiG-23s from Egypt in 1977—the Constant Peg outgrew its corner of the Groom Lake base and made the move, 70 miles north and west, to the remote airfield at Tonopah where hangars and other black-budget-financed infrastructure were soon being built. Just as Groom Lake had its own box of restricted air space in the middle of the Nellis NTTR, Tonopah had its

Tonopah Test Range (TTR), comprising 626 square miles of air space at the northwest corner of the NTTR.

In the meantime, the Red Eagle pilots also had access to other Soviet aircraft, including the fighters of the Sukhoi Design Bureau, which were "owned" by the Air Force Systems Command 6513th Test Squadron. This unit was based at Edwards AFB in California but also conducted operations at Tonopah.

As operations from Tonopah began in July 1979, the Red Eagles were a formal part of Red Flag. Participating pilots who were previously unaware of the top secret existence of the Red Eagles, which included nearly every pilot in the United States armed forces, were each being given an opportunity to fly against a variety of MiGs and to gain an understanding of their flight characteristics. Most American pilots—and only American pilots were permitted to see the MiGs—were duly startled by the exceptional maneuverability of the MiG-21, which had claimed so may American fighters over Vietnam.

Through the end of 1980, the Red Eagles had flown 1,015 sorties, "exposing" 372 air force and navy pilots to the MiGs. Those numbers compared to 87 sorties and 68 exposed pilots at the end of 1979. By 1980, the advanced, variable geometry Mach 2 MiG-23 fighters had been integrated into Constant Peg flight operations under the code name Have Pad.

Two Navy F-14 fighter pilots who had been exposed to the MiGs and to Soviet bloc fighter tactics were Lieutenant Larry Muczynski and Lieutenant James Anderson. In August 1981, over the Gulf of Sidra, off Libya in the Mediterranean Sea, they were attacked by two Soviet-made Su-22s of the Libyan Air Force. Using what they had learned in the skies of the TTR, the Americans counterattacked, destroying both of the Libyan fighters.

Steve Davies, in his excellent book about the 4477th, titled *Red Eagles*, writes at length about the informal and even "outlaw" nature of the 4477th. They operated unacknowledged aircraft with secret budgets and without a formal chain of command that was apparent in the white world. Everything they did, from sourcing parts for maintenance to understanding the flight characteristics of the aircraft, had to essentially be made up as they went, and this task was executed informally, often by word of mouth.

The US Air Force got its first MiG on September 21, 1953, two months after the Korea War ceasefire. It was a MiG-15, hand delivered to Kimpo AB in South Korea by North Korean pilot No Kum-Sok. *USAF*

Efforts made by TAC leadership, including its commander, General Wilbur "Bill" Creech, to require the Red Eagles to conform to established TAC standards and procedures were difficult and often met with resistance. It was like trying to fit Constant Peg into a conventional hole. The TTR was in another world.

The line dividing the worlds was like a Berlin Wall. Few people from the white side were briefed on what happened on the dark side, while those on the dark side could not carry their knowledge of that side into the white world. Secrecy was paramount. The Constant Peg pilots and ground crews could not even tell their families what they did—nor where they did it. Everyone at Tonopah, from pilots to mechanics, had a top secret clearance, and all were confined to the base nearly all the time. Trips into the nearby town of Tonopah were rare.

Above: The first MiG-17 to be evaluated at Groom Lake was a Syrian aircraft obtained from Israel. It was code named Have Drill. *Terry Panopalis collection*

Left: Have Ferry was the second Syrian MiG-17 acquired from Israel. Lieutenant Commander Hugh Brown, US Navy, was lost in the crash of this aircraft in 1979. *Terry Panopalis collection*

Steve Davies discusses the deaths of Navy Lieutenant Commander Hugh "Bandit" Brown and Air Force Captain Mark "Toast" Postai in 1979 and 1982. He tells of the awkward frustration of families who could not know any details beyond the fact that a husband or father would not be coming home and that he had died in the crash of "a specially modified test aircraft."

According to Davies, the most influential and "for many, also the most controversial" of the commanders at the 4477th was Lieutenant Colonel George "G2" Gennin, who assumed command in August 1982. A combat pilot who had flown three tours over Vietnam, he made great strides in tightening up flying and maintenance procedures, especially in the wake of Postai's crash in one of the temperamental MiG-23s. The American pilots who had found the MiG-21 to be maneuverable, responsive, and easy to fly, found the larger, faster, and more complex MiG-23 to be an unforgiving and quirky airplane. It was later learned that Soviet pilots had come to the same conclusion.

Meanwhile, the number of pilots who were exposed to the Red Eagles of the shadowy TTR through Red Flag and otherwise gradually increased through the years from 462 in 1981, to 575 in 1982, and to 666 in 1983.

The first MiG-21 evaluated at Groom Lake flew under the code name Have Doughnut. By 1985, the 4477th TES operated seventeen MiG-21s. *Terry Panopalis collection*

A US Air Force MiG-21 undergoes routine maintenance in a TTR hanger, circa the 1980s. *Terry Panopalis collection*

Through all of these sorties, it was only natural that rumors would seep out, and mention would be made in the media of the MiGs of Tonopah. *Aviation Week* mentioned the secret in its May 18, 1981, issue, reporting erroneously that the US Air Force operated "several squadrons of MiGs," though correctly noting that these aircraft had been obtained from Egypt and that Soviet-built MiG-21s, air defense radar, and electronic warfare equipment were tested against American fighters. By the end of 1983, the Red Eagles fleet consisted of nine MiG-21s and six MiG-23s, with the MiG-17s having been retired.

Little more was said on the outside for several years until a high-profile fatal crash thrust the world of Constant Peg back into the headlines. In its May 7, 1982, issue, *Aviation Week* revealed that Lieutenant General Robert M. "Bobby" Bond, the vice commander of the AFSC, was killed on April 26 "when he tried to eject from what the Air Force said was a 'specially modified test aircraft' during a flight over the Nellis AFB Flight Test Range in Nevada, which started at Tonopah base, 150 miles northwest of Las Vegas. The fact that a three-star general, who would retire in two months, died testing advanced aircraft sparked intense interest, fiercely resisted by the Pentagon."

Above: A MiG-21 flies over the Nellis Range. Despite American markings, MiGs were always a surprise for US Air Force or Navy pilots on their first encounter. *USAF*

Left: Taxiing across the TTR ramp, a MiG-21 heads out to engage American pilots during a Red Flag exercise. *Terry Panopalis collection*

"We will not discuss classified programs," Secretary of Defense Caspar Weinberger said, hoping to keep a lid on the loss of a three-star general in a program that officially did not exist.

Aviation Week speculated that "Bond was flying [either] one of the stealth fighters in a Tonopah program on low radar observable aircraft [an F-117A, which secretly existed at the time], a Soviet MiG-23 fighter, or an even more classified vehicle."

The editors guessed right with the second of their multiple conjectures.

As with any test program, mishaps occurred. This was certainly the case when there were reverse engineered components within aircraft that were unfamiliar—and crudely designed by American standards. The engines, for example, were not up to Western standards. It was surmised that export variants of the MiGs had engines built with planned obsolescence in mind so that replacement engines could be sold for hard currency. There was also the fact that the Soviet ejection seats could have been more trustworthy. In a number of instances, pilots chose to make dead stick landings rather than eject. It was often misgivings about the seat that led to this decision, but the Constant Peg pilots felt a loyalty to the program and were inclined to be protective of its assets.

In Bond's case, as Steve Davies explains, it was an experienced pilot, who was not fully familiar with the MiG-23, who had gotten himself into an irretrievable situation. The Tumansky R-29A engine accelerated beyond Mach 2 in afterburner and would not respond to Bond's attempts to throttle it back. As he lost control of the aircraft, he attempted to eject but broke his neck in the process. The aircraft went down in the southwest part of the NTS.

Personnel of the 4477th TES with one of their MiG-21s inside their hangar at Tonopah.
USAF

During 1984, thanks in part to George Gennin's efforts to make the 4477th function as an operational squadron, sortie rates nearly doubled from the 1983 levels to 2,099. In Red Flag and other exercises, pilots averaged 2.6 exposures to the MiGs compared to 1.8 the year before.

At the end of 1984, the 4477th fleet was comprised of fifteen MiG-21s and nine MiG-23s. A year later, two more MiG-21s and another MiG-23 had been added. In the meantime, it had been decided that the MiGs in the fleet should receive official US Air Force designations. In no small part, the driving force was to give pilots something other than a Soviet "MiG" designation to put in their log book. Indeed, several pilots had flown MiGs hundreds of times. Originally, they had substituted "F-5" for "MiG-21" and "F-4" for "MiG-23," and so on, but this had to change.

It was widely rumored during the 1980s that the MiGs had received triple digit designators, but this has never been officially confirmed. Rumors denied in the 1980s are *still* denied. Nothing official may ever be known. However, numerous sources, from Steve Davies to the meticulous designation chronicler Andreas Parsch, have published the designations.

The final US Air Force fighter designated under the pre-1962 nomenclature system was the F-111, with F-110A being an interim designation for the Air Force F-4 Phantom. For the clandestine designation of MiGs, the US Air Force is said to have returned to the three-digit lineage, using the YF service test prefix. MiG-21 variants were designated as YF-110B through YF-110D, concealing them within the old Phantom designator. The YF-110Cs are reported to have been a dozen Shenyang F-7Bs (the Chinese MiG-21 variant) purchased directly from China in 1987.

During that year, with the F-7Bs added to the 4477th fleet, it was possible to launch as many as eight Red Eagle aircraft for a single "red on blue" encounter during Red Flag operations.

Not all of the other available designation numbers are linked to specific projects. YF-112 is not identified with any certainty, but YF-113 was apparently widely used.

YF-113A was the Have Drill MiG-17; YF-113B was a MiG-23BN; the YF-113C was a Chinese-built Have Privilege MiG-17; and YF-113E was a MiG-23MS (NATO code name Flogger-E). The YF-114C was the Have Ferry MiG-17 (NATO code name Fresco-C); and the YF-114D was the MiG-17 Fresco-D. It is speculated that the YF-116 designation was assigned to an unknown Soviet aircraft, and YF-117 is known to have been assigned to the Lockheed "stealth fighter," also tested at Tonopah during the 1980s and discussed in the following chapter.

Despite the secrecy that still surrounds the MiG operations, the sheer number of blue pilots who had seen or flown against the MiGs made it a hard secret to keep. This was especially true given that part of the reason for the training was for the men who had flown against MiGs to brief other pilots on MiG characteristics and tactics. Those receiving these briefings were naturally suspicious that their briefers had first-hand experience with MiGs *somewhere*.

By 1986 and 1987, in which the 4477th flew 2,792 and 2,793 sorties respectively, the existence of the Red Eagles of Red Flag was an open secret within the US Air Force.

During those years, there had been talk of allowing the 4477th and Constant Peg to be transferred to the white world and for operations to be brought out into the open. Instead, early in 1988, General Robert "Bob" Russ, who had assumed command of the TAC in May 1985, abruptly made the decision to close down the entire operation.

The Red Eagles of the 4477th flew their last mission during Red Flag on March 4, 1988. For this, they launched their largest maximum effort ever, sending thirteen MiG-21s and four MiG-23s against a mixed blue force of F-4s, F-15s, F-16s, and F-111s over the Nellis Range. Captain Mike Scott, the most experienced of all the Red Eagles, credited by Steve Davies with 569 official MiG sorties, was among those in the air that day.

During their nine years, the Red Eagles flew 15,264 sorties against 5,930 blue aircrews. Though operations ended in 1988, the 4477th was not deactivated until 1990, and Constant Peg was not declassified until 2006, fifteen years after the end of the Cold War.

During those fifteen years, Soviet-built— now considered Russian-built—aircraft, including more advanced MiG-29s and Su-27s, have been seen occasionally in the skies over the Nellis Range and have been reported to be operating out of Groom Lake.

General Bill Creech headed TAC from 1978 to 1984, years when the Tonopah MiGs were active. After his death in 2003, Indian Springs Airfield, home of Air Force operational attack drones, was renamed as Creech AFB. *USAF*

General Bobby Bond was a fighter pilot with 213 missions over North Vietnam. In April 1984, as vice commander of the AFSC, he came to Tonopah to fly the MiG-23. Unfamiliar with the aircraft, he lost control and crashed. His high-profile crash fanned the fires of media speculation over the rumors that already circulated about the Tonopah MiGs. *USAF*

CHAPTER 11

FROM HOPELESS DIAMOND TO BLACK JET

ON MAY 5, 1975, Ben Rich was visited in his office by Denys Overholser, a long-time Skunk Works mathematician and radar specialist. Less than four months had passed since Rich had succeeded the legendary Kelly Johnson as head of the Skunk Works, and Rich was already at the crossroads of his career. What the grinning Overholser was about to show him would either be a disaster that would embarrass and humiliate Rich and the Skunk Works or revolutionize warplane technology for the foreseeable future.

"Boss," Overholser said as he handed Rich a drawing of an airplane that looked like an Indian arrowhead. "Meet the Hopeless Diamond."

In his memoirs, Rich described the drawing of an airplane that looked like "a diamond beveled in four directions, creating in essence four triangles."

"How good are your radar cross section numbers on this one?" Rich asked him, as the object of the exercise had been to create an airplane with the lowest possible RCS.

"This shape is one thousand times less visible than the least visible shape previously produced at the Skunk Works," Rich recalls Overholser replying. The radar expert went on to explain that it had an RCS that was much smaller than that of the D-21 drone, which appeared smaller for its actual size on radar than anything the Skunk Works had ever done.

"If we made this shape into a full-size tactical fighter, what would be its equivalent radar signature?" Rich asked. "As big as what . . . a Piper Cub, a T-38 trainer . . . what?"

"Ben, understand, we are talking about a major, major, big-time revolution here," Overholser said. "We are talking infinitesimal."

"What does that mean?" Rich pressed. "On a radar screen it would appear as a . . . what? As big as a condor, an eagle, an owl, a what?"

"Ben," the engineer said with a laugh. "Try as big as an eagle's eyeball."

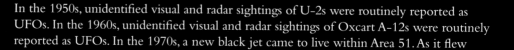

In the 1950s, unidentified visual and radar sightings of U-2s were routinely reported as UFOs. In the 1960s, unidentified visual and radar sightings of Oxcart A-12s were routinely reported as UFOs. In the 1970s, a new black jet came to live within Area 51. As it flew

A pair of Lockheed F-117A Nighthawk stealth fighters. The Nighthawk, known as the Black Jet, was the most talked-about aircraft to come out of the secret world in the Nevada desert since Oxcart. *USAF, Master Sergeant Lance Cheung*

from the runway at Groom Lake, no radar sightings for this aircraft were detected, which embodied what came to be known as "stealth" technology. Like the Angels and Archangels, this airplane began as a secret wrapped in an enigma, which was, in turn, wrapped in the protective walls of the Skunk Works in Burbank, California.

It can be said that the quest for radar cloaking technology began on the first day that an airplane was detected by radar. At the Skunk Works, it began in the late 1950s with the Oxcart program, but no breakthrough was achieved. Through the 1960s, though it was an object of interest at numerous aerospace companies, radar obscuring, or low observables (LO) technology, remained theoretical.

In 1974, the Defense Advanced Research Projects Agency (DARPA) undertook a serious study of LO technology and later commissioned several aerospace companies to develop proposals for accomplishing this elusive magic. The firms included Northrop and McDonnell Douglas – but *not* Lockheed.

Part of the reason for the latter omission was, no doubt, the fact that Lockheed had been teetering on the precipice of going out of business. In 1971, it had taken nearly $200 million in federal loan guarantees to "bail out" the company. Meanwhile, rumors of a substantial series of bribes by Lockheed to foreign government officials were about to explode into an enormous scandal. It was against this backdrop that Ben Rich had succeeded Kelly Johnson in January 1975.

Though the DARPA study was beyond top secret, and Rich was unaware of it, he was already discussing LO technology with Overholser, who thought he might have the key to unlocking the mystery. This key was in a 1962 technical document entitled *Method of Edge Waves in the Physical Theory of Diffraction*, which had been authored by Pyotr Ufimtsev, a Russian engineer at the Moscow Institute of Radio Engineering. He had developed an extremely obscure high-frequency asymptotic theory for predicting the scattering of electromagnetic waves from two-dimensional and three-dimensional objects, which had been read by few and understood by fewer.

As Ben Rich himself noted, "radar cross section calculations were a branch of medieval alchemy as far as the noninitiated were concerned."

He was lucky that Denys Overholser was among the initiates who could comprehend the Ufimtsev Physical Theory of Diffraction (PTD).

"Ufimtsev has shown us how to create computer software to accurately calculate the radar cross section of a given configuration, as long as it's in two dimensions," Overholser explained. "We can break down an airplane into thousands of flat triangular shapes, add up their individual radar signatures, and get a precise total of the radar cross section."

They had to use flat, two-dimensional shapes, because in 1975, computers simply did not have the capacity to model three-dimensional shapes with rounded edges. It was, Rich explains, a matter of "creating a three-dimensional airplane design out of a collection of flat sheets or panels, similar to cutting a diamond into sharp-edged slices.

The Hopeless Diamond evolved into the Have Blue configuration, seen here during RCS pole testing, circa 1976, at the White Sands Ratscat Backscatter Range in New Mexico. *Tony Landis collection*

Amazingly, this all transpired before Rich learned of the secret DARPA project. When he did find out, by accident, he realized that the people at DARPA knew little or nothing about all the top secret work that had gone into the RCS of the A-12 and SR-71. They didn't know about how blending the fuselage and engine nacelles into the wing had served somewhat to reduce the RCS. Nor did they know that the iron ferrite composite materials with which the airframe was coated served to absorb, rather than reflect, radar waves.

Rich lobbied DARPA's director, Dr. George Heilmeier, and finally got the Skunk Works a seat at the table. In August

The Northrup XST was proposed in DARPA's Experimental Survivable Testbed program. This 38-foot scale model is shown here pole tested to evaluate its RCS. *Tony Landis collection*

Above: The first Have Blue prototype underwent engine run-ups at the Skunk Works on November 4, 1977, and was delivered to Groom Lake for flight testing shortly thereafter. *Terry Panopalis collection*

Right: Some sources identify this aircraft as the second Have Blue, HB 1002, but it has the appearance of a full-scale mock-up. *Tony Landis collection*

1975, thanks to the work that was then being done by Overholser and design engineer Dick Sherrer, Lockheed, along with Northrop, made it to the final round to build what was then called the Experimental Survivable Testbed (XST).

Within the Skunk Works, the Hopeless Diamond concept had many detractors, not the least of which was Kelly Johnson himself, who still came in from time to time on a consulting basis. For Ben Rich, the turning point came on September 14, 1975, when a Diamond mock-up proved itself to have an RCS that was one tenth of one percent of that of the Tagboard D-21.

Evaluation of the radar-evading properties of the Hopeless Diamond configuration began in December 1975 using a 38-foot scale model on a pole. The results, conducted at various facilities from the California desert to the White Sands of New Mexico, were good. But they were *so* good that Ben Rich's bosses, Lockheed President Larry Kitchen and CEO Roy Anderson doubted that they could be replicated on a full-sized airplane.

"We've been as good as our predictions up to now," Rich told Kitchen. "There's no reason to think we'll drop the ball."

Meanwhile, because the Diamond had been designed mainly for its LO characteristics, the aircraft was inherently very unstable. Nothing quite like it had ever flown. Many people, including Kelly Johnson, doubted it *could* fly, even if it was constructed with quadruple-redundant fly-by-wire controls.

Nevertheless, Ben Rich persevered, and in April 1976, Lockheed received the nod to build the first prototype, which would be code named Have Blue. By this time, the program had been transferred from DARPA to the AFSC. Though not officially acknowledged, the "Have" prefix in code names is believed to have indicated an AFSC program.

Northrop meanwhile continued to develop its own rendering of stealth technology under DARPA's Battlefield Surveillance Aircraft, Experimental (BSAX) program. This undertaking evolved into the advanced technology bomber (ATB) and ultimately into the B-2 Spirit "stealth bomber." This aircraft is as remarkable as those included in this book but is not profiled because its gestation and testing took place at Palmdale and Edwards AFB and not within the confines of Area 51 and the Nellis Range.

The Skunk Works built two Have Blue demonstrator aircraft at Burbank, which were designated as Articles HB 1001 and HB 1002. Like the Hopeless Diamond, the Have Blue craft were sharply angular, had flat, faceted surfaces, and resembled arrowheads. They were 47 feet 3 inches long and 7 feet 6.25 inches high at the tail. The wingspan was 22.5 feet, but most noticeable was the 72.5-degree sweep of the modified delta wings. The engine was the relatively small General Electric J85-GE-4A turbojet, delivering 2,950 pounds of thrust. It was first ground-tested in HB 1001 at Burbank on November 4, 1977.

Twelve days after the engine run-up tests, HB 1001 was partially disassembled and airlifted to Groom Lake in a C-5 transport. The first flight took place on the first day of December, with Lockheed pilot Bill Park, a veteran of the Oxcart program and a man well-known around Area 51, in the cockpit. Park demonstrated that the unusual configuration would fly, and he made thirty-five successful flights.

Ultimately, the undoing of HB 1001 was not its unorthodox configuration, but its landing gear, which was borrowed from the thoroughly vetted F-5 fighter. On May 4, 1978, Park made a hard landing and lifted off to come around for a second try. This time the gear would not extend properly, and the engine flamed out as he was trying to gain altitude to eject. Park survived his ejection but suffered injuries which permanently grounded him.

HB 1002 made its debut flight on July 20, 1978, with Ken Dyson, who had served as chase pilot for the HB 1001 flights, at the controls. Over the course of the ensuing year, HB 1002 made fifty-two flights and was thoroughly tested against numerous radar threats. These were considered to have been successfully completed by the time HB 1002 was lost. A hydraulic leak resulted in the aircraft becoming uncontrollable, and Dyson was forced to eject.

Nevertheless, the US Air Force remained enthusiastic about the promise of the new "stealth" technology, and on November 1, 1978, they issued a contract for a larger, operational aircraft under the code name Senior Trend. As noted earlier, the "Senior" prefix in two-word code names is believed to indicate a program that was of direct interest to US Air Force Headquarters.

The designation assigned to the aircraft was F-117, following in the series of numbers allegedly assigned to the

This view of Have Blue HB 1002 in flight looks almost unreal. Many people doubted that the faceted, dart-like airframe could fly, much less evade radar. It did both. *Tony Landis collection*

Tonopah MiGs. The five service test aircraft were designated as YF-117; the subsequent fifty-four production series birds, F-117A. The name Nighthawk was later assigned to the aircraft, although the early test aircraft were dubbed Scorpions, and aircrews typically referred to the F-117 as the Black Jet. The latter was as much literal as metaphorical, as the operational aircraft were, in fact, painted black. Another nickname, "Wobblin' Goblin" (or "Wobbly Goblin"), a reference to the inherent instability of the Have Blue prototype, was widely mentioned in the media, but not used by aircrews.

The F-117 was much larger than Have Blue and somewhat different in configuration. For example, the vertical tail surfaces had been canted inward on Have Blue and were redesigned as a "V" tail in the F-117. The latter was 65 feet 11 inches long and 12 feet 9.5 inches high at the tail. The wingspan was 43 feet 4 inches, and the sweep of the wings was 67.5 degrees.

Though the F-117 was referred to as a stealth *fighter*, it was armed with neither guns nor air-to-air missiles. It was to be strictly a ground attack aircraft. The offensive ordnance, a variety of precision-guided munitions, was carried internally, because to hang it externally would defeat the stealth characteristics of the airframe. The signature weapon of the F-117 would be the 2,000-pound GBU-27 Paveway III precision-guided "smart" bomb, of which two were carried.

The Senior Trend Black Jet was powered by a pair of General Electric F404 turbofan engines, of the same family as the engines powering the F/A-18 Hornet strike fighter, each of which delivered 10,600 pounds of thrust. The aircraft would have a service ceiling of 45,000 feet and an unrefueled range in excess of 1,000 miles, although it was designed for aerial refueling.

Like Senior Crown (SR-71) and Senior Bowl (D-21) before it, Senior Trend was cloaked in secrecy of very high classification. As with Have Blue, the aircraft would be towed into hangers whenever Soviet spy satellites were known to be passing over Groom Lake.

However, as with Lyndon Johnson in 1964, President Jimmy Carter was running for reelection in 1980 and eager to share his prodefense credentials with voters. As Johnson knew of Senior Crown, Carter knew of Senior Trend, and he was willing to lift the veil of secrecy. The Carter administration was more surreptitious, leaking the news anonymously. During the week of August 10, the media was abuzz with news of "stealth technology," as items appeared in *Aviation Week & Space Technology* and the *Washington Post*, as well as on *ABC News*.

Secretary of Defense Harold Brown held a press conference on August 22 to clarify the leak, admitting that the limited details previously published were accurate. In an article entitled "Stealth—an Invisible Aircraft with High Political Visibility," published in the *Christian Science Monitor* on September 5, Peter Stuart pointed out that the Carter revelation would backfire by allowing the opposition to claim that military secrets were divulged for political purposes. Two months later, Carter lost the election.

The fear was that the Soviet Union would undertake a program of countermeasures. However, they apparently

Pole testing the F-117 production configuration at the Lockheed's facility near Helendale, California in the Mojave Desert. *Lockheed*

Left: The first YF-117A takes shape, circa 1978. *Lockheed*

Below: The F-117 series differed from the Have Blue prototypes in size and in the "V" tail surfaces being angled outward rather than inward. *Lockheed*

did not take stealth technology seriously—despite the fact that their own Dr. Ufimtsev had achieved the original breakthrough. When Ufimtsev came to the United States in 1990, he met with Ben Rich, telling him that "senior Soviet designers were absolutely uninterested in my theories."

After Brown's press conference, secrecy once again closed around Senior Trend. No pictures were released—nor would they be for eight years—and most analysts predicted an aircraft with rounded, not faceted, surfaces. In that sense, the secret remained safe. Nor was the F-117A designation made public. During the 1980s, it was generally assumed that the stealth fighter would be designated F-19.

The first flight of the YF-117A (tail number 79-10780) came on June 18, 1981, with Lockheed pilot Hal Farley at the controls. Additional YF-117A aircraft joined the flight test program by early 1982. The first aircraft were painted light gray or in camouflage colors similar to the markings of the Have Blue, but operational aircraft were painted black.

In October 1979, the US Air Force had activated the 4450th Tactical Group to take delivery of the F-117As as they became operational. Based at Nellis AFB, the 4450th was first formed as a unit flying A-7D Corsair II attack bombers. This formation was in part to provide a cover story for the unit and in part to provide pilot training in an aircraft of a similar size. The A-7Ds would also serve as chase planes during the test program.

As this program had been unfolding, the air force was engaged in a major upgrade of the airfield at the TTR, where the Red Eagle MiGs operated, and which was to be the home of the new stealth fighter. New hangars and other facilities were added, and the runway was lengthened from 6,000 to 12,000 feet. In May 1982, the 4450th relocated to the TTR under the command of Colonel James Allen. In the early years, tanker support for the 4450th was provided by the 9th SRW, which had considerable experience supporting black projects such as Senior Crown.

The first loss of an F-117 and its pilot came on July 11, 1986, when Major Ross Mulhare went down over a remote area of Southern California. The cover story that it was an A-7D crash was belied by the heavy security around the crash site. The media, especially *Aviation Week*, speculated that a "stealth fighter" had crashed, but there was no official confirmation. More than two years later, on November 10, one week after the 1988 election and after years of speculation, the F-117 was officially revealed to the press and public. A year later, in October 1989, the 4450th was inactivated and its activities were assumed by the reactivated 37th Tactical Fighter Wing (TFW), commanded by Colonel Tony Tolin. It was not until April 1990 in an event at Nellis AFB that F-117s were first displayed in a public air show. The fifty-ninth and last Nighthawk was delivered in July 1990.

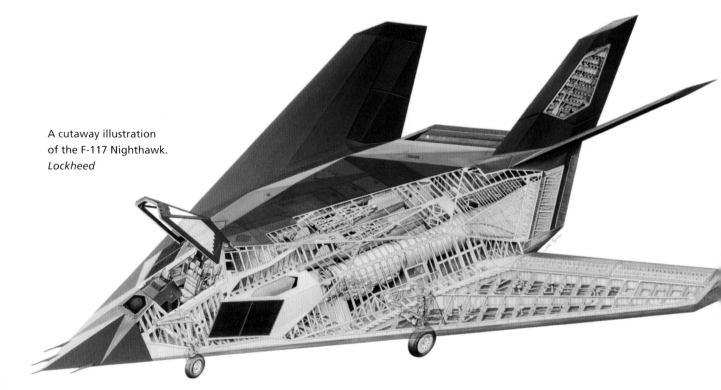

A cutaway illustration of the F-117 Nighthawk. *Lockheed*

In the meantime, F–117s had flown their first combat mission in December 1989 during Operation Just Cause, which heralded the removal of Panamanian dictator Manuel Noriega. Using multiple aerial refueling, six Nighthawks flew from Tonopah to Panama, though only two were authorized to attack and their strikes were purely diversionary. Given the inadequacies of Panamanian radar, Just Cause was hardly a fitting first test of the stealth technology of the Black Jets, but a more suitable opportunity would soon be afforded.

Six months after Just Cause, in August 1990, under the command of Colonel Alton Whitley, three squadrons of the 37th TFW made their first overseas deployment to King Khalid AB at Khamis Mushait in Saudi Arabia. This deployment came as part Operation Desert Shield, the buildup of forces in the region in response to Saddam Hussein's invasion of Kuwait. Because Khamis Mushait is located in a high desert environment reminiscent of Nevada, it came to be known to the 37th TFW crews as "Tonopah East."

The first twenty-two aircraft were from the 415th Tactical Fighter Squadron (TFS), known as the "Nightstalkers," commanded by Colonel Ralph Getchell. They were followed in December by eighteen aircraft from the 416th TFS, the "Ghostriders," and a month later by a small number of F–117s from the 417th Tactical Fighter Training Squadron (TFTS).

After several months of preparation and war gaming, the F–117s were tasked with being the first aircraft to strike downtown Baghdad as Desert Shield became Operation Desert Storm. Given that the Iraqi capital had some of the best radar, SAM, and antiaircraft defenses in the world, there was still considerable trepidation that Nighthawk losses would be high.

On the night of January 16, 1991, as the air war began, F–117s were the tip of the spear, striking a series of high-value targets across Baghdad. The sky was filled with a wall of defensive fire but devoid of radar returns from the stealth fighters. The 37th TFW emerged unscathed.

Top: Two F-117As wait at twilight with the Tonopah control tower in the background. The Nighthawks moved here from Area 51 in 1982. *Lockheed*

Bottom: The Nighthawk hangar complex at Tonopah. In 1989, the 37th TFW was activated here. *Lockheed*

"The Black Jet was able to operate with impunity over Iraq, being the only machine allowed into the heavily defended skies over downtown Baghdad," recalled Lieutenant Colonel Barry Horne of the 415th TFS. "They dropped the first bombs of the war, hitting the central communications building of the Iraqi military machine. . . . Our guiding principle was that we concentrated on high-value, heavily defended targets, which lent themselves to the use of precision-guided munitions. That means we went after aircraft hangars, command and control bunkers, telecommunications, power plants in the early stages, line-of-communication targets like bridges. . . . Later, it was footage from their imaging infrared weapons system which showed the world the amazing accuracy of modern precision-guided weapons."

Over the ensuing six weeks of Desert Storm, but mainly in the early days of combat, the F-117 dropped more than 2,000 tons of precision-guided munitions, mainly GBU–10s and GBU–27s.

"Combat missions could last up to six hours," Horne explains. "It would start the day before, getting the tasking mid-to-late evening. Mission planning people would look at the task, coordinating with other agencies, like air defense, but most importantly with the tankers. That took quite a time, working out times and locations that we would need refueling, and just how much fuel we'd need. . . . We'd plan to launch in daylight, late afternoon, wanting to use as much of the night over the target as possible. We might fly two or three waves a night, depending on how far away the targets were. Some targets were close together, and we'd go in a package, while others were widespread."

134

In 1997, in the General Accounting Office (GAO) report *Operation Desert Storm: Evaluation of the Air Campaign*, the US Air Force noted that "the F-117 was the only airplane that the planners dared risk over downtown Baghdad."

The US Air Force reckons that although the Nighthawks represented only 2.5 percent of the engaged combat aircraft, they succeeded in hitting 31 percent of the total targets. They flew 1,271 sorties in 6,900 hours, with a mission-capable rate of 85.8 percent—better than when the aircraft was operating at Tonopah.

The F-117, once a shadowy program discussed only in whispers, emerged from the Gulf War victory as a media celebrity. The Black Jet, and its ability to deliver "smart bombs" with both impunity and precision, was the very symbol of American technological primacy in weapons of combat.

However, the once-secret weapon emerged from the shadows into a changing world. All sixty-five of the originally planned F-117s had been delivered, but talk of reopening production ensued. Talk emerged of a navalized "F-117N Seahawk" that would operate from aircraft carriers and of Britain's Royal Air Force buying an F-117C variant—but neither materialized.

By the end of 1991, the red flag with the hammer and sickle had come down from above the Kremlin for the last time. Both the Soviet Union and the Cold War were now things of the past. So too, in the minds of many whose fingers were on the budgetary purse stings in Washington, was the need for the Nighthawk. This was the era of the "Peace Dividend," when defense was a rapidly shriveling priority for the United States and when the US Air Force was halving its number of tactical wings.

The US Senate voted to acquire an additional two dozen Nighthawks for the US Air Force, but this vote was later overturned as lawmakers looked forward toward the

The hypothetical carrier-based F-117N was considered by the US Navy. *Lockheed*

F-117As of the 37th TFW at Langley AFB, Virginia, are shown en route to Saudi Arabia during Operation Desert Shield in 1990. *USAF, Master Sergeant Boyd Belcher*

next-generation F-22. Another Lockheed product, the F-22 was a high-performance fighter that embodied stealth characteristics. Those characteristics made buying more Nighthawks seem less desirable. No more would ever be built.

To be fair, the delicate faceted surfaces of the F-117 were difficult to maintain. While the F-117 had been designed in an era of slide rules and primitive computers, the F-22 benefited greatly from quantum leaps in computer-aided design technology that had been made in the intervening years.

As the Black Jets came out of the metaphorical black world, they departed from the black world of Area 51 and Tonopah. In May 1992, the Nighthawk wing began the move to Holloman AFB in New Mexico and from the 37th TFW to the 49th TFW. The latter, formerly an F-15 unit, had a lineage going back to World War II when America's top scoring ace of all time, Richard "Dick" Bong, flew with the unit in the Southwest Pacific.

Through the coming years, the 49th TFW deployed overseas numerous times for activities including Operation Southern Watch over Iraq in the late 1990s and Operation Allied Force over the former Yugoslavia in 1999. It was during the latter that the Nighthawk fleet suffered its only combat loss. On March 27, an F-117 piloted by Lieutenant Colonel Dale Zelko was acquired by Serbian radar as its bomb bay doors opened. Zelko escaped, but the downed aircraft was recovered, and its secrets were shared with the Russians and the Chinese.

Some years later, a Russian surface-to-air missile engineer told David Fulghum of *Aviation Week* that because the airframe was badly damaged "we haven't been able to model the entire [low-observable bomber]. It's not the same as testing against an undamaged F-117. You provide us with a complete stealth aircraft and then we'll tell you how effective we are."

The Nighthawks subsequently flew missions during Operation Enduring Freedom in 2001 and Operation Iraqi Freedom two years later while deployed with the 379th Air Expeditionary Wing. In the early hours of March 19, 2003, F-117s flew the first mission of the war in Iraq, an attack against a bunker in a little-used presidential palace on the edge of Baghdad where Saddam Hussein was erroneously rumored to be hiding.

This was the opening salvo of the Black Jet's last war. The Nighthawk came home from this war to face its own retirement. Essentially, the US Air Force needed to free up funds for the costly and high priority F-22 program. In 2005, it was decided that over a billion dollars could be saved by moving the planned retirement of the F-117 back from 2011 to 2008. In March 2007, the process began, with the first of the aircraft being flown back to Tonopah from Holloman.

The last operational Nighthawk was retired in August 2008, although reports of F-117s still flying over the Nellis Range persist. Unlike the previous birds of Area 51, such as the A-12s and SR-71s, only a handful of developmental Nighthawk models have been relegated to serve as museum pieces. Since their retirement, more than forty of the original aircraft are maintained in climate-controlled facilities in the continuing security of their original Tonopah home.

Top left: Back home to the Nellis Range, an F-117A from the 49th FW refuels during a 2002 Red Flag exercise. The Black Jets were retired five years later. *USAF, Senior Airman James May*

Top right: Sergeant David Owings, crew chief, communicates with Major Joe Bowley, a pilot with the 37th TFW, during a preflight check in Saudi Arabia in 1991. *USAF, Sergeant Kimberly Yearyean*

CHAPTER 12

AURORA, BLACK MANTA, AND THINGS THAT WENT PULSE IN THE NIGHT

THE TRUTH THAT many secrets lie within Area 51 is probably the only fact about the place that is not a secret. Secrecy breeds speculation, and at Area 51, this has swirled for decades around the question of "what else is out there that 'they' aren't telling us about?"

To catalog all the rumors is a losing task. On one end of the spectrum are stories of levitation, time travel, and extraterrestrials. On the more practical side, the black airplanes that have been revealed invite speculation about those that have not. By the 1980s, the growing current of interest in secret black airplanes was surging in parallel with the long-running current of interest in flying saucers.

There has naturally been conjecture about a faster and higher flying reconnaissance aircraft. Two such programs that have been talked about since the days when the A-12 and SR-71 were new are Isinglass and Rheinberry. Gregory Pedlow and Donald Welzenbach, who wrote the CIA's official history of Aquatone and Oxcart, report that the CIA "Office of Special Activities did briefly consider several possible successors to the Oxcart during the mid-1960s. The first of these, known as Project Isinglass, was prepared by General Dynamics to utilize technology developed for its Convair Division's earlier FISH proposal . . . in order to create an aircraft capable of Mach 4-5 at 100,000 feet. General Dynamics completed its feasibility study in the fall of 1964, and OSA took no further action because the proposed aircraft would still be vulnerable to existing Soviet countermeasures."

They go on to discuss Rheinberry, describing it as "a more ambitious design from McDonnell Aircraft" that had appeared in 1965. As they explain, "This proposal featured a rocket-powered aircraft that would be launched from a B-52 mother ship and ultimately reach speeds as high as Mach 20 and altitudes of up to 200,000 feet. Because building this aircraft would have involved tremendous technical challenges and correspondingly high

This illustration of the mythic Aurora aircraft takes into account descriptions of many witnesses who saw an unidentified delta-winged aircraft during the 1980s and 1990s. *Illustration by Erik Simonsen*

costs, the Agency was not willing to embark on such a program at a time when the main emphasis in overhead reconnaissance had shifted from aircraft to satellites."

One should fast forward to vehicles flying in the twenty-first century. Discussed in chapter 14, these include the X-51A Waverider and the Hypersonic Technology Vehicle 2 (HTV-2). The latter is designed for Mach 20 speeds.

In the 1990s and beyond, the signature unseen black airplane of Area 51 was Aurora. Indeed, the very term "Aurora" has stood for everything that those who haunt the periphery of Area 51 yearn for. It was, and still is, the very archetype of the high performance black aircraft whose secrecy the government will do anything to maintain.

Otherworldly craft? No, lenticular clouds. However, *these* were actually photographed over Area 51. In the annals of Project Blue Book, such clouds were often reported as "flying saucers."
Bill Yenne

When Aurora was first mentioned in *Aviation Week and Space Technology* in 1989 and discussed at length in its October 1, 1990, issue, the floodgates of rumor opened. After the recently exposed Lockheed F-117 stealth fighter and Northrop B-2 stealth bomber, people were anxious for further revelations from the dark world of dark airplanes. Described at the time as "a possible hypersonic reconnaissance aircraft" that was the worthy successor of the recently retired SR-71, Aurora was the right black airplane at the right time.

Even the means by which Aurora came to light are like a scene from 1990s technothriller. As *Aviation Week* recalls, "A line item identified as 'Aurora' in a Fiscal 1986 Procurement Program document dated Feb. 4, 1985, supposedly was simply one 'site' for B-2 bomber funds when that program was highly classified, according to a government official. Listed under the 'Other Aircraft' category, 'Aurora' was projected to receive sharply increased funding. The Fiscal 1986 budget request for Aurora—$80.1 million—jumped to $2.272 billion in Fiscal 1987, according to the document. Because it was listed under a strategic reconnaissance section, the Aurora reference was widely thought to be a subtle admission that an SR-71 replacement was under development. However, given the B-2 funding practices of the time, the Aurora name may not have been related to a new hypersonic aircraft as originally believed."

Then, in 1989, an engineer named Chris Gibson, working on the *Galveston Key* oil platform in the North Sea, had reported an unidentified triangular aircraft being refueled by a KC-135 Stratotanker. Given that Gibson had served with the Royal Observer Corps and was, as Britain's *Guardian* newspaper put it, "an expert on recognizing aircraft," his account was given "considerable credence." His sighting was later associated with the Aurora legend.

Bob Lazar, who had been propelled to international media attention through his reports of extraterrestrial beings in the possession of the US government, added his own recollections. He now reported that he had *also* seen Aurora aircraft operations while he was inside Area 51.

In 1990, *Aviation Week* prophetically reported that "advanced secret aircraft developed at highly classified government facilities in the Nevada desert over the last decade are demonstrating and validating new technologies for the US's future fighters, bombers and reconnaissance platforms. Although facilities in remote areas of the Southwest have been home to classified vehicles for decades, the number and sophistication of new aircraft appear to have increased sharply over the last 10 years, when substantial funding was made available for 'deep black' projects."

Aviation Week wrote of "a high-speed aircraft characterized by a very loud, deep, rumbling roar reminiscent of heavy-lift rockets. When observed at medium altitude, this aircraft type often makes a pulsing sound and leaves a thick, segmented smoke trail or contrail. Lighting patterns indicate the aircraft is on the order of 100+ feet long, but no reliable description of a planform has been reported."

In southern California and elsewhere, numerous reports of these unusual pulsing sounds in the sky were recorded, as well as many sightings of contrails described as

The hangar complex inside Area 51 looks as Bob Lazar might have known it during the time that he reported seeing alien spacecraft. Here, normally dry Groom Lake can be seen with water in it after a rare, heavy rain. *Tony Landis collection*

resembling "doughnuts on a rope." These were mostly attributed to Aurora, though another delta-winged aircraft, known variously as "Brilliant Buzzard" or "The Mothership," were also mentioned in some of the conspiracy theories about Area 51.

In 1994, in his memoir, *Skunk Works*, Lockheed's Ben Rich more or less confirmed Aurora's connection to the B-2, which had emerged from the black world in 1988, shortly before Aurora entered the lexicon of pop culture.

"The funding for the competition [to create a long-range stealth bomber] came out of a secret stash in the Air Force budget," Rich writes. "A young colonel working in the Air Force 'black program' office at the Pentagon, named Buz Carpenter, arbitrarily assigned the funding the code name Aurora. Somehow this name leaked out during congressional appropriations hearings, the media picked up the Aurora item in the budget, and the rumor surfaced that it was a top secret project assigned to the Skunk Works—to build America's first hypersonic airplane. That story persists [to 1994] even though Aurora was the code name for the B-2 competition funding. Although I expect few in the media to believe me, there is no code name for the hypersonic plane, because it simply does not exist."

Nevertheless, aviation journalists and aviation enthusiasts had already seized upon Aurora as an emblem for the black aircraft of the era, and there was no turning back.

Aurora, as articulated in the 1990s, may or may not have existed, but, given how little has been revealed about Aurora and the extremely high performance successors to the Oxcart and its family, these stories about the deep black world seem as fresh today as they did at the time. Indeed, to this day many black airplane enthusiasts believe that Aurora is still "out there."

Area 51, that mysterious other world or parallel universe, might be better identified as R-4808N. This restricted air space includes not only the area around Groom Lake called Dreamland (top right), but also most of the Nevada Test Range (left) where nuclear weapons and nuclear propulsion schemes were extensively tested. *FAA*

One area where the interests of the black airplane and flying saucer enthusiasts has intertwined is in the suggestion that the flying saucers actually *were* black airplanes, disc-shaped craft created here on earth. Back in 1947, General Nathan Twining had mentioned a "possibility that these objects are of domestic origin—the product of some high security project not known to [the Headquarters Air Staff]."

If so, what were they?

One tangent that has been followed is that of secret German advanced aeronautical projects, whose files and hardware were scooped up after World War II and brought to the United States. Germany had fielded the world's first operational jet fighter, the Me 262; the first jet bomber, the Arado Ar 234; and the first rocket-propelled interceptor, the Me 163. They also had Wernher von Braun's V-2 ballistic missile and the foundations of the technology that would put humans into outer space.

As with revelations about secrets within Area 51 that invited speculation about "what else is out there," the revealed advances in German aeronautics begged the same question.

Among the myriad projects in wartime Germany was the Horten IX, a tailless flying wing created by the brothers Walter and Reimer Horten. This aircraft was built by Gothaer Waggonfabrik under the designation Go 229, and made its first powered flight in February 1945. Like most of the German experiments, the Horten aircraft was largely forgotten over time. That is, until 1988, when the US Air Force rolled out the B-2, which was a flying wing with no vertical tail surfaces. There was now a renewed interest in the "what else?" of wartime German secrets.

Among the most unusual "what else" aircraft languishing (or said to be) among the old German designs were those of various *disc-shaped* aircraft. These aircraft include ones developed by an aeronautical engineering group that included Dr. Richard Miethe and Rudolf Schreiver, who discussed them in an interview published by the magazine *Der Spiegel* in 1950. In his book *German Jet Genesis*, David Masters mentions a disc-shaped aircraft created by Schreiver and Miethe with a diameter of 138 feet, which climbed to nearly 40,000 feet in three minutes during a February 1945 test flight. Could this aircraft have been secretly imported to the Nevada desert?

The idea that flying saucers were real, and that they were created in Nazi Germany, is the sort of sensational hypothesis that thrills conspiracy theorists, but no concrete evidence has ever emerged to confirm these tantalizing rumors. On the other hand, they *could be* another "high security project not known" outside Area 51.

This US Geological Survey view of the Groom Lake complex was taken when Have Blue was here. It has yet to be proven—or disproven—that Aurora and Black Manta were lurking down there. *Tony Landis collection*

In 1989, an unidentified triangular aircraft, later identified as Aurora, was seen refueling from a KC-135 over the North Sea. Here is a similar view closer to Area 51. *Illustration by Erik Simonsen*

A contemporary of Aurora during the late 1980s and early 1990s was a black airplane that was known as the "TR-3A Black Manta," though no one seems to know whether this appellation was officially assigned or simply the product of the imagination of an outside observer. A possible source for the name is in DC Comics, where the name Black Manta had first been used in 1967 to identify a villain who was an enemy of the comic book icon Aquaman.

The assumption behind the TR-3A designation is that this aircraft was a tactical reconnaissance aircraft numbered in the same lineage as the TR-1 (later redesignated as U-2R) and the NASA variant of that aircraft designated as ER-2 (for Earth Resources). Reports that were officially denied by the US Air Force suggest that the Black Manta was secretly used during Operation Desert Storm in early 1991, possibly to laser designate targets for F-117A strikes.

However, while the pictures of Aurora that have been published can be characterized as something between hypothetical and pure fantasy, some basis of fact for the supposed configuration of the TR-3A Black Manta depictions appeared in the 1990s. These were remarkably similar in appearance to drawings that were submitted to the US Patent Office in 1975 by Teledyne Ryan.

This company was no stranger to the world of black airplanes and aircraft built secretly for clandestine CIA operations. At exactly the same time that Kelly Johnson's Skunk Works was creating Aquatone and Archangel, aircraft that officially did not exist except under undisclosed CIA contracts, Teledyne Ryan was providing the CIA with a family of high-performance, remotely piloted vehicles (RPVs) called Lightning Bugs.

A component of Northrop Grumman since 1999, Teledyne Ryan traced its roots to the Ryan Aeronautical Company of San Diego, which was acquired by Teledyne in 1969. The firm was founded by T. Claude Ryan, whose claim to fame was the Ryan NYP, better known to the world as Charles Lindbergh's *Spirit of St. Louis*. Having built more than a thousand primary trainers during World War II, Ryan turned to jet-propelled target drones after the war. Beginning in 1951, the company became successful with its long line of shark-nosed Q-2 Firebee drones.

Ryan proposed Firebees for reconnaissance missions as early as 1959, and after the high profile losses of manned U-2 and RB-47 spyplanes over the Soviet Union in 1960, the armed forces were interested. Under the designation BQM-34, Ryan produced numerous multimission and reconnaissance Firebees for use by the US Air Force and the US Navy during the Vietnam era. The CIA also made extensive use of the reconnaissance variant under the nonmilitary designation Model 147A and the official name Lightning Bug. Just as the CIA's Oxcart Archangels had been shadowy ghosts of the Air Force SR-71 Blackbirds, the CIA's Lightning Bugs were like shadow-world ghosts of the Firebees.

In the meantime, for deep penetration missions into Chinese airspace beyond the range of the Firebee and too dangerous for a manned U-2, the US Air Force initiated the Compass Arrow project. This resulted in the Teledyne Ryan Model 154 Firefly, a long-range, remotely piloted vehicle with a fuselage similar to that of the Firebee but with the engine atop the fuselage to minimize detection from the ground and with long wings reminiscent of those of the U-2. Its fuselage was also coated with radar-absorbing materials. Designated as AQM-91A, the Firefly first flew under a veil of secrecy in 1968 and was acknowledged only after a 1969 crash. Though a number of Fireflies were built, they are not known to have been used operationally over China by the CIA or the US Air Force.

Teledyne Ryan later adapted its Compass Arrow/Firefly technology for the Compass Cope program. Initiated in 1971, this project envisioned an RPV roughly analogous in range and performance to the U-2. The two aircraft were the Boeing Model 901, which made its first flight in 1973 as the YQM-94A Gull (Compass Cope-B), and the Teledyne Ryan Model 235, which debuted in 1974 as the YQM-98A Tern (Compass Cope-R). In 1976, the Gull was chosen over the Tern for a production contract, but the US Air Force cancelled the Compass Cope project before any production aircraft were built.

This timeline of the covert world of Teledyne Ryan reconnaissance aircraft in the 1970s leads directly to the DARPA 1974 study of radar-evading LO technology. By the time Lockheed and Northrop were refining the F-117 and the B-2 respectively, down the coast in San Diego, Teledyne Ryan was also conducting its own LO research. In the final weeks before August 1975, when Lockheed and Northrop were awarded DARPA study contracts under the XST program, Teledyne Ryan made two filings with the US Patent Office. These are the aircraft often considered to be the tangible predecessors, even prototypes, of the Black Manta.

The first of the Teledyne Ryan patent applications was filed on July 21, 1975; the second came just three weeks later on August 14. Like the Firebee, Firefly, and Tern, the first aircraft, known by Patent Number 4019699, was an unmanned RPV described as an "Aircraft of Low Visibility." The second, identified as Design 244265, was a manned aircraft roughly the same size as the Skunk Works Have Blue. Both aircraft had a triangular, delta planform, and both were like the Firefly and Tern in that they had their engines atop the fuselage. The configuration of Design 244265 was very similar to the sketches of the Black Manta, which appeared in the aviation media in the 1990s.

By the time the two patents were granted, on April 26 and May 10, 1977, Lockheed had designed but not yet test flown the stealthy Have Blue.

The 4019699 was described as possessing "very low observability visually and to radar, thermal and acoustic detecting devices. The aircraft is designed to have as few edges and surfaces as possible, such as a delta wing type, the edges being straight, or near straight

Little is known of the mysterious AQM-91A Firefly, the "grandfather" of the Black Manta. *Author's collection*

and the vertices rounded. All surfaces are as near flat as possible, within the limits of aerodynamic requirements, and the entire surface of the aircraft is electrically conductive with minimum discontinuity. The propulsion unit is faired into and shielded from radar by the wing. Portions of the aircraft, particularly edges, may be of radar absorbing material, and any essential openings, ducts or protruding fins are similarly shielded to minimize the radar cross section."

Listed on the patent documents as the "inventors" of 4019699 were Teledyne Ryan engineers Robert Wintersdorff and George Cota, while Waldo Virgil Opfer was credited as the inventor of Design 244265. Opfer had previously established himself in the field of vertical take-off and landing (VTOL) technology. In 1967, he received Patent Number 3333793 for a balanced lift yaw control system for VTOL aircraft, and the following year, he and John M. Peterson were awarded Patent Number 3388878 for a VTOL aircraft with a balanced power, retractable-lift fan system. Essentially they had designed a system whereby a Lockheed F-104 Starfighter could take off and land vertically like a British Aerospace Harrier.

Despite Opfer's credentials, there is no indication that the Design 244265 aircraft was capable of VTOL operations. However, UFO sightings that occurred in Belgium in 1989 and 1990 describe a triangular vehicle that was capable both of hovering and of high speed flight. At the time, some conspiracy theorists suggested a link between these sightings and the triangular shape of the notional Black Manta.

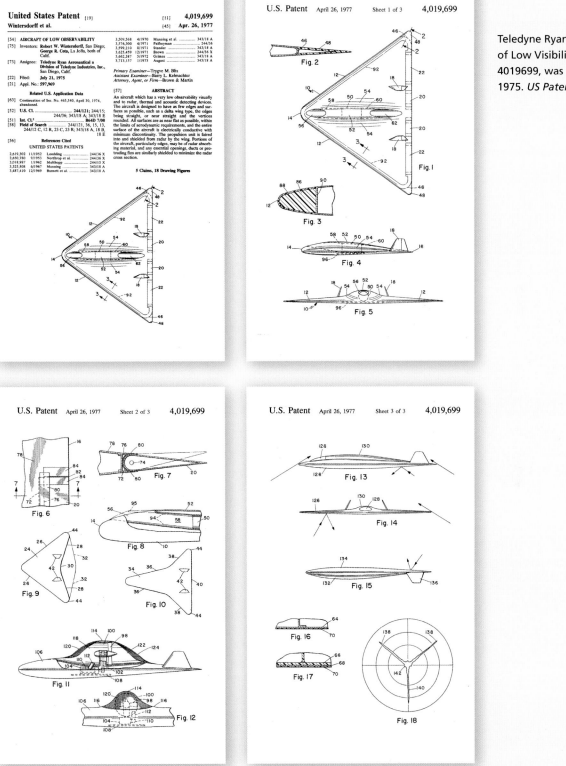

Teledyne Ryan's Aircraft of Low Visibility, Patent 4019699, was filed in July 1975. *US Patent Office*

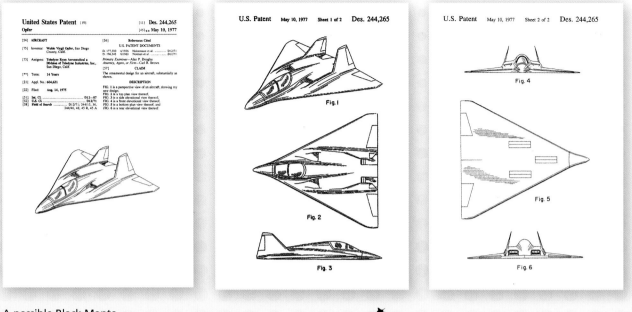

A possible Black Manta precursor was Teledyne Ryan Aircraft Design Number 244265, filed in August 1975.
US Patent Office

While little has been reported about the Black Manta since the turn of the century, Aurora still crops up occasionally in media accounts of strange rumblings and inexplicable tremors that occur from time to time across Southern California and the desert southwest.

One stolid and credible believer is aviation journalist Bill Sweetman, who has written extensively about black airplanes and their world and about Aurora in the aerospace media, including *Aviation Week* and *Jane's Defence Review*. He finds continuing evidence for the existence of Aurora not in the night sky, but back where it was originally discovered—in the federal budget.

In *Popular Science* magazine in June 2006, Sweetman wrote that "my investigations continue to turn up evidence that suggests current activity. For example, having spent years sifting through military budgets, tracking untraceable dollars and code names, I learned how to sort out where money was going. This year [2006], when I looked at the Air Force operations budget in detail, I found a $9 billion black hole that seems a perfect fit for a project like Aurora. Over the years, I've learned that few people investigate budget holes seriously."

Steven Kosiak of the Center for Strategic and Budgetary Assessments, a Washington, DC, think tank, told Sweetman that "a fair amount of classified spending goes through in supplemental requests. It's seen as must-pass legislation, and people don't look at it closely."

An obvious question is why Aurora, if it exists, had been in flight test for over a quarter century. Sweetman answers this question by explaining that "the main hold-up has probably been fuel. The way to make a hypersonic cruiser work is to use circulating fuel to soak up the engine's heat, but conventional jet fuel can't absorb enough heat to do the job. In the 1980s, Aurora would have been designed to use fuels such as hydrogen or methane, which are gaseous at normal temperatures and had to be supercooled and densified to fuel the aircraft. Although that strategy is possible, it's not operationally easy, and complicated refueling would be counterproductive for a jet intended to provide prompt overflight when the military needed it. Better fuels and engine technologies exist now."

If 244265 or 4019699 was Black Manta's precursor, perhaps this tailless, ray-like UAV is its successor. *Illustration by Erik Simonsen*

It is not certain whether all of the mysterious sights and sounds described in the same sentence with Aurora really are Aurora, whether they are other secret airplanes, or whether they are even attributable to the same source. Nor is it certain that the "$9 billion black hole" that Bill Sweetman reported in 2006 was actually Aurora, something else, or many something elses.

Aurora may be unveiled tomorrow, or it may soon be revealed that Aurora once existed but that it was scrapped and buried in the gravel of Area 51 several decades ago.

As of this writing, it has not been seen. Nor has the Black Manta, nor any of many other mysterious rumored aircraft known only from glints in the sky or unexplained reverberations. This does not mean that there is nothing to the reports, only that there has been nothing tangible—*yet*.

CHAPTER 13

WHALES, BIRDS OF PREY, AND THE TRUE DEEP BLACK

DURING THE YEARS when the black airplane enthusiast community was abuzz with rumors and rumblings about Aurora and Black Manta, other aircraft were secretly flying over Area 51, airplanes that were ultimately confirmed to have existed. Among them was also one well-known aircraft that officially never flew, although rumors persisted that a variant of this aircraft had been seen.

Back around the late 1980s and early 1990s, those in the black airplane and UFO communities generated a lot of talk about triangular aircraft, some of which were linked to Aurora. Behind all the innuendo, one triangular aircraft actually did exist. It was a stealth aircraft developed for the US Navy that was given the attack designation A-12 —coincidentally the same designation that Lockheed had given the Oxcart built for the CIA. Nicknamed "the Flying Dorito" because of its shape, the A-12 is not believed to have ever flown, but reports indicate that a subscale demonstrator aircraft had been seen over the Nellis Range.

The origins of this flying wing attack aircraft go back to the early 1980s. At that time, the US Air Force was starting to define its twenty-first century fighter requirements under the Advanced Tactical Fighter (ATF) program. This ultimately led that service to a fly-off between the Lockheed YF-22 Lightning II (later Raptor) and the Northrop YF-23 Black Widow II (also Gray Ghost). Meanwhile, in 1983, the US Navy's Naval Air Systems Command (Navair) was doing the same with its need for a new generation advanced tactical aircraft (ATA). The ATA was defined as an all-weather, long-range, low observable (LO), high-payload, carrier-based attack aircraft that was to have replaced the Grumman A-6.

Unlike the ATF program, which was widely discussed in the media, the US Navy's ATA project was kept secret until 1986, ostensibly because of its stealth nature. Even then, Navair revealed as little data as it could. For example, the Navair phone directory listed Captain Sam Sayers as ATA program coordinator but didn't list an address.

In November 1986, two teams—Grumman/Northrop (which merged in 1994) and McDonnell Douglas/General Dynamics—were selected to participate in the demonstration and evaluation phase of the ATA project. In October 1987, the latter team

Known affectionately as the Whale, the Northrop Tacit Blue Low Observable technology demonstrator made more than 100 flights through Area 51 air space between 1982 and 1985 without any conspiracy theorist guessing that it existed. *Illustration by Erik Simonsen*

was chosen to build the aircraft, which was designated A-12 in 1988 and later named Avenger II. It was initially reported that the US Navy might acquire as many as 620 A-12s, while the US Marine Corps desired 238.

However, the program was cursed by delays and cost overruns, which ultimately led to its cancellation in January 1991 by Secretary of Defense Dick Cheney. All that was left of the aircraft were the lawsuits, which made their way through the courts for years.

Even as the triangular aircraft, which may or may not have been there, was being reported over the Nellis Range, nobody noticed an airplane that *was* there but looked like a flying milk carton. Northrop's Tacit Blue is remarkable for the fact that it flew its entire flight test program without ever having been the subject of revelations in the media or speculation among enthusiasts.

Over a period of three years, from 1982 to 1985, Tacit Blue made more than a hundred flights in broad daylight, apparently without being seen by outsiders. It then landed at Groom Lake, taxied into a hangar, and was never heard from again for years. It was not known outside Area 51 until eleven years after its last flight. It might well have *never* been known to the outside world if the US Air Force had not deliberately declassified it in 1996.

Tacit Blue is also noteworthy for its appearance. Through the early 1990s, the golden age of Aurora speculation, the pages of the enthusiast magazines were filled with artist's conceptions of black world aircraft that looked like saucers, arrows, darts, and Doritos. They had the appearance that one might call sleek, swift, sinister, or all of the above. Tacit Blue, on the other hand, had the appearance of a large box with some stubby, straight wings clumsily attached. Even its friends called Tacit Blue "the Whale."

Like Lockheed's Have Blue, Northrop's Tacit Blue simply did not have the appearance that fit the 1980s layman's idea of what a LO stealth airplane should look like, but the engineers knew different. The Northrop aerodynamicists were as proud of their Whale as the people at the Skunk Works were of their Hopeless Diamond.

According to Peter Grier, who wrote about Tacit Blue for *Air Force Magazine* in 1996, just after the declassification, "The California headquarters for Northrop's advanced projects division was filled with pictures of whales. There were whale paintings in the lobby, whale drawings on letterheads, and whale logos stamped on all kinds of company equipment. Northrop employees sometimes referred to each other as 'whalers.' Top managers had models of whales on their desks."

Visitors who were not in the know were left to assume that it was someone's idea of an interior decorating scheme.

The Tacit Blue program got underway in 1978, several years after Have Blue began. Some of the Northrop engineers who had worked on DARPA's XST program—which lost out to Have Blue—transferred to Tacit Blue.

Like Have Blue, Tacit Blue originated with DARPA, and it later passed to the US Air Force. While the idea behind Have Blue was to strike heavily protected, high-value targets with precision weapons, Tacit Blue was part of a program designed to counter a massive Soviet ground attack in central Europe. A component of the larger Assault Breaker program, the Whale's role was to be an undetectable aerial command post that could operate behind enemy lines and provide real-time targeting information to battlefield commanders. Under Assault Breaker, the attacking armies would be subjected to death from above by huge numbers of precision munitions, guided and directed by the Whale.

A similar program, which emerged at around the same time and would ultimately benefit from technology developed under Tacit Blue, was the Joint Surveillance Target Attack Radar System (Joint STARS). This program also centered on an airborne platform for battle management and command and control specifically aimed at detecting and

Developed by McDonnell Douglas and General Dynamics in the late 1980s, the triangular A-12 Avenger II attack aircraft was the very image of a mysterious Area 51 black airplane. In fact, its troubled life and dramatic demise were scrutinized in the media. *Author's collection*

targeting enemy ground forces behind
enemy lines. Joint STARS evolved from
separate US Army and Air Force programs
that were merged in 1982. The system was
developed by Grumman Aerospace, though
the aerial platform, later designated as
E-8A, was a Boeing 707 airframe.

Above: The flight
deck of the Tacit
Blue demonstrator
aircraft. *USAF*

Left: Just a handful of
inflight photographs
of the Tacit Blue were
released in 1996 when
the curtain was pulled
off the program. *USAF*

The unusual and ungainly shape of
Tacit Blue derived from a case of form
following function, not an attempt to make
fun of the black airplane buffs. As Grier
points out, Northrop engineers discovered
that "the radar reduction needs of a surveillance aircraft were turning out to be far more
demanding than those of a bomber or strike fighter. The latter two types of aircraft
generally fly straight toward targets and defending radars and then turn and fly away. The
theory behind their designs was to minimize the radar returns of their front and rear views.
Tacit Blue's concept of operations, however, called for it to loiter behind enemy lines while
flying in circles. It would be exposed to detection devices operating on all sides."

Another engineering problem was that of "shoehorning the radar's big antenna into a
relatively small aircraft, while making the antenna's field of view large enough."

John Cashen, a Northrop veteran of the XST who was an engineer on Tacit Blue,
explained that "integrating the antenna created the boxy nature of the body. The rest of the
design was driven around trying to get this box to fly and to make it [low observable from]
all aspect[s]."

As Grier points out, Northrop did not use the same faceted surface methodology as
the Skunk Works had with Have Blue. Rather, they used curvilinear, or Gaussian, surfaces
"to redistribute a radar beam's electrical energy." This same approach to aircraft surfaces
was also applied to Northrop's B-2 program, which was already underway. This curvilinear

Tacit Blue is shown parked on a taxiway at Wright-Patterson AFB as it arrived at the National Museum of the US Air Force. *USAF*

A good front view of the once-mysterious Tacit Blue aircraft. *USAF*

The "V" tail surfaces of the Northrop Tacit Blue are canted outward, as are those of the Lockheed F-117A Nighthawk. *USAF*

Northrop's "V-tailed," diamond-winged YF-23A Gray Ghost was an ATF contender that pioneered a unique aerodynamic configuration. *Author's collection*

approach led to a configuration that defied radar detection, but it was also a configuration that John Cashen described as "an aircraft that at the time was arguably the most unstable aircraft man had ever flown."

Grier goes on to say that the Whale was so "unstable in both pitch and yaw, it depended on a quadruple-redundant, General Electric fly-by-wire control system for safety. . . . If a model of Tacit Blue was balanced on its point of gravity and placed in a wind tunnel, it would weathervane around until it was pointing tail first into the onrushing wind. The airplane's nose did not have to be pulled up very much before the whole thing would threaten to flip over."

It was 55 feet 10 inches long and 10 feet 7 inches high, with a wingspan of 48 feet 2 inches. It weighed 30,000 pounds and was powered by a pair of Garrett ATF3-6 high-bypass turbofan engines. Unlike the Joint STARS airborne radar platform, which carried a crew of radar operators, the Whale was a single-seat aircraft in which the radar was to be operated by a ground station, as is the case in more recent unmanned reconnaissance drones.

Beginning in February 1982, the Tacit Blue aircraft made 135 flights, often logging multiple missions in a day, and racked up 250 hours. Despite the lone Whale's notorious aerodynamic instability, it survived to its retirement.

Northrop had started work on two Tacit Blue airframes at the company's Hawthorne, California plant, but only one was completed. The second remained unfinished as a backup, which could have been completed as needed but was not. Neither was there a production Tacit Blue variant, because its mission could be better performed by the more conventional Joint STARS E-8A. As Grier explains, "Joint STARS, with its large twenty-nine-foot-long antenna and superior depth of view, could perform much of Tacit Blue's mission by itself. Compounding the problem for the Whale was that Joint STARS cost less, had longer endurance, was air refuelable, and could scan a wider area."

General George Muellner, who commanded the Air Force's 6513th Test Squadron during Tacit Blue flight testing and who was later deputy assistant secretary of the US Air Force for acquisition, explains that the Tacit Blue program "turned into a test-bed because its low-observable technologies proved to be more valuable than its [mission] contribution."

He further explained that the Gaussian stealth was one of the most important breakthroughs in defense technology in the latter half of the twentieth century. "The B-2

Like the CIA Lightning Bugs over Vietnam in the 1960s, the Senior Prom vehicle was airdropped from a Lockheed DC-130 mothership. *Tony Landis collection*

Left: The airframe of the Lockheed Skunk Works Senior Prom vehicle was based on Have Blue. *Tony Landis collection*

Right: A Senior Prom vehicle shows high visibility markings, its wings folded back. *Tony Landis collection*

exploits it, as does the F-22 and a lot of other vehicles," he said. What he did not mention is that the same technology was utilized by Northrop in the development of its YF-23A demonstrator, which lost out to the Lockheed YF-22A in the ATF fly-off. Indeed, the YF-23A's butterfly tail and engine configuration were reminiscent of the Whale's.

At the official unveiling of the Whale on April 30, 1996, Muellner remarked that the Tacit Blue RCS was "below that of a bat, somewhere down in that area. I haven't looked at the RCS of a bee recently."

Even as the Whale and the details of its flight test program were revealed to the public, another Area 51 black program from the same era remained in the shadows, still classified, and known only through whispers. A Lockheed Skunk Works product, Senior Prom was a stealthy cruise missile demonstrator that was based on the faceted airframe of Have Blue. It made its first flight at Groom Lake in October 1978, placing it seven years ahead of the General Dynamics stealthy Advanced Cruise Missile (ACM), which entered service with the US Air Force as the AGM-129A.

During the flight test program, Senior Prom vehicles were air dropped over the Nellis Range by DC-130 motherships operating out of Groom Lake. Though Senior Prom was reported to have been cancelled in 1982, the program remains classified long after Tacit Blue was brought out of the black world. Since the beginning of the twenty-first century, rumors have circulated that the still-secret Senior Prom has been used as a reconnaissance drone.

In 1998, two years after disclosing Tacit Blue, George Muellner retired from the US Air Force and took a job as head of Phantom Works. This is the McDonnell Douglas advanced research and development unit that became part of Boeing in 1997 when the two companies merged. Muellner later served as president of advanced systems for Boeing's Integrated Defense Systems business unit.

The Phantom Works were formed in 1991 at Lambert Field near St. Louis and named for the F-4 Phantom, the most successful warplane ever produced by McDonnell before its merger with Douglas. As the name suggests, the Phantom Works were the McDonnell Douglas answer to the Lockheed Skunk Works. Mark Gottschalk, the Western Technical Editor of *Design News,* wrote in 1996 that comparisons to the "legendary Skunk Works are inevitable," but he went on to quote Gerry Ennis, vice president of the Phantom Works Prototype Center, who explained, "We not only develop new technologies and processes, we demonstrate them and then move them to production."

One of the first projects undertaken by the Phantom Works was NASA's X-36 Tailless Fighter Agility Research Aircraft. As described by NASA's Dryden Research Center, "It was designed to fly without the traditional tail surfaces common on most aircraft. Instead, a canard forward of the wing is utilized, in addition to split ailerons and an advanced thrust-vectoring nozzle for directional control. The X-36 is unstable in both the pitch and yaw axes; therefore, an advanced, single-channel digital fly-by-wire control system, developed with some commercially available components, stabilizes the aircraft."

Beginning in 1994, Phantom Works built two X-36 vehicles that were 19-foot, 28-percent-scale models of a potential fighter. They were remotely controlled by a pilot in a virtual cockpit at a ground station. The X-36 made its first flight in May 1997 and completed thirty-one successful flights through November. The entire flight test program took place over Edwards AFB, rather than the Nellis Range, but if a full-sized variant existed, or exists in the black world or in the future, it is not impossible that its flight test program might be based out of Groom Lake.

The unmanned X-36 Tailless Fighter Agility Research Aircraft, seen here in a photo from October 1997, was one of the first ventures from the Phantom Works. *NASA, Carla Thomas*

One of the most intriguing products of the Phantom Works that really *was* a denizen of Area 51 on George Muellner's watch was the Bird of Prey. Visually, this aircraft possessed the strange but sleek design characteristics that are appreciated by those in the world of extraterrestrial buffs. It was even named after an alien spaceship. According to James Wallace, writing in the *Seattle Post-Intelligencer* of October 18, 2002, it is named for a class of Klingon starships first seen in the 1984 motion picture *Star Trek III: The Search for Spock*.

The Bird of Prey project began at the Phantom Works in 1992, and the aircraft made its first flight eight months earlier than the X-36, on September 11, 1996. Unlike the publicly acknowledged NASA X-36, the Bird of Prey was a classified program, funded by the contractor but apparently managed by the US Air Force. Like Tacit Blue, it was successfully obscured from public view until its flight test program concluded in April 1999 after forty successful flights.

The Bird of Prey program, like many of the projects that have come out of the Phantom Works, utilized rapid prototyping techniques to cut both costs and development time. According to the company, the program "pioneered breakthrough low-observable technologies and revolutionized aircraft design, development and production." The Bird of Prey was also one of the first aircraft programs to "initiate the use of large, single-piece composite structures; low-cost, disposable tooling; and 3D virtual reality design and assembly processes to ensure the aircraft was affordable to build as well as high-performing."

The aircraft was not, however, one to push the edge of the performance envelope. The Bird of Prey had a reported cruising speed of 300 mph and a modest service ceiling of 20,000 feet. The aircraft was 47 feet long and had a wingspan of about 23 feet. It weighed about 7,400 pounds and was powered by an off-the-shelf Pratt & Whitney JT15D-5C turbofan engine.

The man who is considered to have been the "father of the Bird of Prey" was Alan Wiechman, the director of signature design and applications for the Phantom Works whose career in LO design had begun at the Lockheed Skunk Works, where he worked on Have Blue and the F-117 program, as well as the *Sea Shadow*, Lockheed's stealth warship.

The Bird of Prey first flew in 1996 but was not officially unveiled until 2002. *Tony Landis collection*

In April 2002, six months before the Bird of Prey was officially declassified, Wiechman received the 2001 Technical Achievement Award from the National Defense Industrial Association (NDIA) for his work in LO aircraft design. The NDIA called him "a giant whose work to date has given the United States a legacy of improved survivability and influenced an entire generation of combat vehicles," adding that "because of Wiechman's work, the United States gained a 15-year lead over potential adversaries that it has not relinquished, and the effectiveness of his designs and products has been thoroughly proven in combat operations."

On October 18, 2002, the Bird of Prey was made public. According to Boeing, the

The Boeing Bird of Prey inflight. Its flight test program reportedly lasted from 1996 to 1999. *Author's collection*

Below left: The Bird of Prey glides over Groom Lake on a hazy afternoon. *Tony Landis collection*

reason was that "the technologies and capabilities developed [in the program] have become industry standards, and it is no longer necessary to conceal the aircraft's existence."

Jim Albaugh, president of Boeing Integrated Defense Systems, proudly bragged that with the Bird of Prey, the Phantom Works "changed the rules on how to design and build an aircraft."

George Muellner, a man with two decades of experience in black world airplanes and now an executive with Integrated Defense Systems, added that the Bird of Prey program "is one of many that we are using to define the future of aerospace."

The fact that Boeing itself funded the project to the tune of $67 million hints that "the future of aerospace" involves secrets, more intriguing than the Bird of Prey, that have been in the sky over Area 51 since before 2002.

They may be revealed tomorrow, or they may remain mysteries indefinitely.

Above right: The archetypal denizen of Area 51, the Bird of Prey was never detected by the black airplane enthusiast community during its flight test program. *Illustration by Erik Simonsen*

CHAPTER 14
SEND IN THE DRONES, WATCH FOR THE BEASTS

IN THE WORLD of popular interest in aviation, the first decade of the twenty-first century could be called the Decade of the Drone. Certainly the drone has become synonymous with clandestine military operations in places such as Afghanistan and Pakistan. These aircraft, called RPVs in the third quarter of the last century and unmanned aerial vehicles (UAV) more recently, were first widely used for aerial reconnaissance missions during the war in Southeast Asia and for combat operations since the turn of the century.

However, drones have been around for almost a century. The first military drone designed for combat, the Kettering Bug, was being tested in 1918 and might have played a role in World War I had that conflict not ended when it did. By midcentury, the hobby of flying remotely controlled airplanes was widespread. During World War II, an enterprising Californian named Reginald Denny started a company called Radioplane, which sold thousands of scaled-up radio-controlled models to the US Army and US Navy as target drones. Northrop later bought Radioplane and continued building both piston-engine and jet target drones.

After World War II, as discussed in chapter 12, Ryan Aeronautical (later Teledyne Ryan) developed a line of jet-propelled target drones that were called Firebees. In turn, these were adapted for reconnaissance missions over North Vietnam, Laos, and elsewhere during the 1960s, propelling the company down a path toward more sophisticated RPV reconnaissance aircraft such as Compass, Compass Cope, and Patent Number 4019699.

By the 1990s, as RPVs were now being called UAVs, they were growing in sophistication. Many surveillance drones in service then, as now, such as the RQ-2 Pioneer, RQ-5 Hunter, or Boeing Insitu ScanEagle, were small, slow, piston-engine aircraft. Others are larger and much more sophisticated. Teledyne Ryan's experience with the Compass Cope, for example, led to the RQ-4 Global Hawk, a large jet aircraft that can operate at 60,000 feet and stay aloft for more than 24 hours.

Drones captured the headlines after 2001, not only for their reconnaissance capability, but also for their offensive capabilities. Arming the General Atomics RQ-1 Predator with Hellfire missiles to take out terrorists and insurgents came as an afterthought to its original

The Lockheed Martin P-175 Polecat was a worthy successor to the reconnaissance aircraft projects that had come to Groom Lake from the Skunk Works through the decades. *Illustration by Erik Simonsen*

concept. However, when operations were watching the enemy in real-time video feeds, there was a natural inclination to attack the enemy that could be seen. The reconnaissance RQ-1 became the armed, multimission MQ-1, and General Atomics developed the much larger and more capable MQ-9 Reaper, which was capable of carrying more and varied offensive ordnance.

In the meantime, in 1998, DARPA and the US Air Force had initiated the Unmanned Combat Air Vehicle (UCAV) program, aimed at demonstrating unmanned stealth aircraft that could be flown on deep penetration missions, such as suppression of enemy air defenses, into heavily defended air space. This led to the Boeing Phantom Works X-45A, first flown in 2002, and the Northrop Grumman X-47A Pegasus, which made its debut in 2003.

The future of aerial combat predicted at the turn of the century was embodied by the X-45A UCAV prototype, seen here at the Dryden Flight Research Center in 2001. *DARPA*

Under the program, the Phantom Works project utilized technology from the Bird of Prey, while the X-47A would have benefited from the obscure Teledyne Ryan 4019699 research.

The UCAV program later became the Joint Unmanned Combat Air System (J-UCAS). The "J" was inserted when the US Navy expressed keen interest in the previously US Air Force program, specifically in the X-47A. Meanwhile, UCAV became UCAS when the DOD decided to consider the overall system developed within a program, not just the vehicle. Meanwhile the term UAS was introduced to supersede UAV, although in practice the UAV acronym continued to prevail.

The "white world" X-45A and X-47A never inhabited that arcane world where their existence was denied and are not known to have been tested in the skies over Area 51. However both projects involved technology that suggests that they may have "black world" cousins whose existence may not be disclosed for decades.

One unmanned aircraft definitely straddles that line in the Lockheed Martin RQ-3 DarkStar, which brings the narrative full circle and back to the Skunk Works. Indeed, the DarkStar has evolved into programs that are known to have been seen over Area 51 and to others that simply remain unknown.

The DarkStar originated as Lockheed's entry in DARPA's High Altitude Endurance (HAE) UAV advanced airborne reconnaissance program, which was initiated in 1993 and which DARPA managed on behalf of the Defense Airborne Reconnaissance Office (DARO). HAE had evolved out of DARPA's High-Altitude, Long-Endurance (HALE) program of the 1980s that led to the development of the Boeing Condor, a huge UAV with a service ceiling of 67,000 feet.

During the 1990s, each of the armed services developed a complicated and confusing taxonomy of "tiers" to define its UAV program, but this practice fell out of use in later years, probably because the respective nomenclature of the services did not align, and within the services, the drones themselves did not conform precisely to the tiers. The two complementary US Air Force aircraft developed under HAE were described not as Tier II and Tier III, but as Tier II Plus and Tier III Minus. Tier II Plus identified a strategic UAV operating up to 65,000 feet with a range of 3,000 miles, slightly beyond the performance envelope of Tier II. Tier III Minus UAVs were strategic HAE vehicles embodying LO characteristics, but they had a shorter endurance than Tier II Plus or Tier III aircraft.

Under HAE, the Tier II Plus aircraft was the RQ-4 Global Hawk, while the Tier III Minus was the RQ-3 DarkStar. Both were intended to be reconnaissance aircraft, not UCAVs. The DarkStar first flew in 1996, while the RQ-4 Global Hawk first flew in 1998. The Global Hawk was a longer endurance, higher flying aircraft, while the DarkStar was designed as a stealth aircraft capable of operating in hostile environments.

While elements of the Global Hawk's ongoing twenty-first century operational career remain classified, it has been widely photographed and was never a black program. The DarkStar, meanwhile, came and went quickly, leaving much speculation about follow-on aircraft.

Resembling a "flying saucer" from the front, the DarkStar airframe was composed primarily of nonmetal composites, and it had no vertical tail surfaces. It was only 15 feet from front to back, but its wing spanned 69 feet. The DarkStar had a gross weight of 8,600 pounds and was powered by a single Williams FJ-44-1 turbofan engine. It had an endurance of 12.7 hours, or eight hours above 45,000 feet.

The first DarkStar prototype made a successful debut flight in March 1996 at Edwards AFB but crashed on its second flight a month later. After twenty-six months of reworking, the second DarkStar made a forty-four-minute fully autonomous first flight in June 1998, but the Defense Department officially terminated the Tier III Minus program in January 1999. By this time, it seemed that there was more interest in the potential usefulness of a long range Global Hawk than a stealthy DarkStar.

The first X-45A, with its weapons bay door open, dropped its first bomb in March 2004. *DARPA*

However, the Skunk Works was already at work on a successor to the RQ-3. Reports denied at the time spoke of a "Son of DarkStar" that saw action during Operation Iraqi Freedom in 2003, but just two years after that, a Skunk Works flying wing UAV *was* flying in the skies over Area 51.

This aircraft, which could possibly be considered a "nephew of DarkStar," was developed by Lockheed Martin as Unmanned System P-175 and named Polecat. This is a term that describes a member of the weasel family but is also synonymous with "skunk" in American slang, therefore suggesting a reference to the Skunk Works. Polecat could also be a reference to the pole that is used to hold a model of an aircraft aloft when evaluating the RCS of its airframe shape. Whatever

the origin of the moniker, the P–175 was first flown in 2005 and disclosed to the media in July 2006 at the Farnborough International Air Show in England. The aircraft itself made no appearance, remaining safely beyond public gaze within the confines of Area 51.

Lockheed Martin announced that it had developed the new aircraft to demonstrate that the Skunk Works still had the "right stuff" to compete in the advanced technology world where the principal players had been Boeing and Northrop Grumman. Frank Mauro, the company's director of unmanned systems, told Amy Butler of *Aviation Week and Space Technology* that the company undertook the project against the backdrop of a perception within the industry that it had abandoned unmanned aerial vehicle technology after DarkStar. "We've taken some hard shots in the past three or four years that [we were]

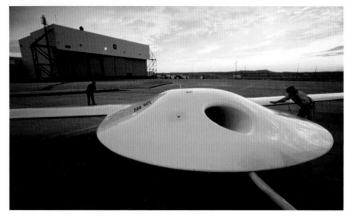

Above: When UCAV became J-UCAS, the X-45A went away, but the other J-UCAS aircraft remained. December 2012 found the X-47B aboard the aircraft carrier USS *Harry S. Truman* in the Atlantic. *US Navy*

Left: Before Polecat, there was the RQ-3 DarkStar, seen here being prepped for a dawn patrol. *Lockheed Martin*

Above: The P-175 Polecat demonstrator over Groom Lake, circa 2006. *Lockheed Martin*

Left: An RQ-3 DarkStar on the flight line, circa the late 1990s. *Lockheed Martin*

not in the UAS game," he said, "and there is a perception that our future is at risk. We are putting our money where our mouth is."

Indeed, the company had spent $27 million of its own money on the program, which Mauro described as a "significant" proportion of the company's research aircraft budget during the period. However, it is also significant that the Polecat was developed in a year and a half, a very short time to bring an aircraft that involves innovative technology from initial concept to first flight.

Only one Polecat was built, constructed of 98 percent composite materials, and with a wingspan of 90 feet. The company acknowledged having pioneered a low-temperature curing process for composites used in the aircraft. In this case, the composites were cured at 150 degrees Fahrenheit rather that the 350 degrees of a conventional autoclave. The idea was cost savings. The Polecat had a gross weight of 9,000 pounds and was designed with a payload bay between the wings that could accommodate a half ton of sensors,

Above: An artist's conception of an RQ-170 Sentinel in flight. *Bill Yenne*

Right: The 30th Reconnaissance Squadron was reactivated in 2005 at Tonopah to fly the RQ-170. *USAF*

reconnaissance gear, or weapons. It was powered by two FJ44-3E Williams International engines.

Frank Cappuccio, the executive vice president and general manager of Advanced Development Programs and Strategic Planning at Lockheed Martin later said that "no one has ever developed in this configuration a high lift-to-drag ratio, and we are going to do it higher than anyone has done it. . . . [The Polecat] was specifically designed to verify three things: new, cost-effective rapid prototyping and manufacturing techniques of composite materials; projected aerodynamic performance required for sustained high altitude operations; and flight autonomy attributes. In addition, the company investment and the resulting successful flights are proof positive of our commitment to developing the next inflection point in unmanned systems."

He added that the engine intakes were masked to deflect radar and explained that without vertical structures and a tail, the aircraft was "inherently low-observable" though it had not been "coated" with radar-reflecting material because "it is not expected to fly operationally."

An innovative "twisting strut" inside the Polecat's wings had, according to Lockheed Martin, been designed to "flex in air and improve the laminar flow over its swept wings, propelling the UAV to high altitudes." The intended operational altitude of the Polecat was specified at 60,000 feet, much greater than that of the UCAV/UCAS demonstrators.

An important design feature that the DarkStar and Polecat shared with the X-45 and the X-47, as well as with the B-2, is the absence of vertical tail surfaces. This design feature, which helps to reduce the RCS, is still considered to be very leading edge technology in the twenty-first century. However, recall that this innovation had been pioneered a half century earlier in Germany by Walter and Reimer Horten.

Lockheed Martin intended to continue flying the Polecat in an ambitious series of high-altitude test flights, but on December 18, 2006, over the Nellis Range, the sole Polecat prototype suffered what Lockheed Martin characterized as an "irreversible unintentional failure in the flight termination ground equipment, which caused the aircraft's automatic fail-safe flight termination mode to activate." The aircraft was destroyed in the ensuing crash.

In March 2007, *Flight International* reported that the notion of "building a replacement" for the Polecat was under consideration.

When a Lockheed Martin statement affirmed that a Polecaat replacement was "certainly being discussed," few outside the Skunk Works and the shadowy corners of the world of Tonopah and Groom Lake realized that something else was already in development. Nor was there any indication that such a thing might be afoot until a tailless flying wing very similar to the Polecat was observed in the skies over Afghanistan later in 2007.

This aircraft was Lockheed Martin's Sentinel, a UAV that bore the designation RQ-170. Assigned for no reason that was then apparent, this designation was far out of sequence with the US Air Force reconnaissance drone nomenclature that then topped out at just eighteen with the Boeing YMQ-18A Hummingbird, a rotorcraft UAV. The higher number suggests that the RQ-170 designation was that used by the CIA, where manufacturer model numbers, rather than military designations, are used. The Lockheed A-12 and the Ryan Model 147 are examples of such aircraft. The Polecat did bear the nearby company designator, P-175. It is possible that Sentinels were operated by both the US Air Force and the CIA, as are Predator drones.

The Air Force Sentinels were based inside the deep black world of the Nellis Range and the Tonopah Test Range, where they were assigned to the 30th Reconnaissance Squadron. This unit dated back to World War II and was operational until inactivated in 1976. Reactivated in 2005, the 30th was assigned first to the 57th Operations Group at Nellis AFB and then to the 432nd Air Expeditionary Wing at Creech AFB, the umbrella organization for UAVs, such as the Predator and Reaper, that were active over southwest Asia.

Before its existence and its designation were officially acknowledged in 2009, the RQ-170 was glimpsed as it operated out of Kandahar AB in Afghanistan. For want of an official name or designation, the mysterious UAV came to be known as the "Beast of Kandahar."

The revelation to the media, also identifying Lockheed Martin as the manufacturer, came on December 4, 2009, after which little further information was released. It was stated that the US Air Force was "developing a stealthy unmanned aircraft system (UAS) to provide reconnaissance and surveillance support to forward deployed combat forces. . . . The fielding of the RQ-170 aligns with Secretary of Defense Robert M. Gates's request for increased intelligence, surveillance and reconnaissance (ISR) support to the Combatant Commanders and Air Force Chief of Staff General Norton Schwartz's vision for an increased US Air Force reliance on unmanned aircraft."

Writing in *Aviation Week*, David Fulghum and Bill Sweetman observed, "Visible details that suggest a moderate degree of stealth (including a blunt

In December 2011, this RQ-170 landed or crashed with minimal damage, in Iran. *Image released by the Iranian government*

leading edge, simple nozzle and overwing sensor pods) suggests that the Sentinel is a tactical, operations-oriented platform and not a strategic intelligence-gathering design. Many questions remain about the aircraft's use. If it is a high-altitude aircraft it is painted an unusual color—medium grey overall, like Predator or Reaper, rather then the dark gray or overall black that provides the best concealment at very high altitudes. . . . The wingspan appears to be about 65 feet, about the same as an MQ-9 Reaper. With only a few images to judge from, all taken from the left side, the impression is of a rather deep, fat centerbody blended into the outer wings. With its low-observable design, the aircraft could be useful for flying the borders of Iran and peering into China, India and Pakistan for useful data about missile tests and telemetry, as well as gathering signals and multi-spectral intelligence."

Two months later, in February 2010, Bill Sweetman reported in *Aviation Week* that an unidentified aircraft matching the description of the Beast, had been seen over Korea. He wrote, "The Beast of Kandahar gets around. The hitherto-classified Lockheed Martin RQ-170 Sentinel unmanned air vehicle (UAV), its existence disclosed after our inquiries in December, has been sighted outside Afghanistan. A Korean newspaper report—overlooked when it appeared in December—has now surfaced and states that the UAV had been flying for several months from a South Korean base—probably Osan, where the US Air Force currently operates U-2s—before it was disclosed. This revelation points directly to an answer to one of the puzzling questions about the Beast: why would you use a stealthy aircraft to spy on the Taliban? The answer is that you don't, but Afghanistan and South Korea have a common feature: they are next door to nations with missile development programs."

In August 2010, David Fulghum reported that "the latest twist is that the US Air Force's stealthy RQ-170 Sentinel flying wing either has returned or is returning to operations in Afghanistan, this time with a full motion video (FMV) capability that ground commanders have been demanding as part of the continuing ISR buildup in the country. What's not clear is whether the Sentinel's stealth enables the conduct of unobserved surveillance missions near or over the borders with Iran and Pakistan."

A KC-135R refuels a twenty-first century stealth UAV with a configuration similar to the Polecat's. *Illustration by Erik Simonsen*

A hypothetical UCAS aircraft from the futuristic rookery at the Skunk Works launches a Hellfire missile. *Lockheed Martin*

It was no secret that American UAVs had been operating routinely over Pakistan for years, and Sentinels were probably also active. Numerous RQ-170 missions were reportedly flown in preparation for Operation Neptune Spear, the successful effort to take out Osama bin Laden, who was killed at his compound in Abbotabad, Pakistan, on May 2, 2011, by US Navy SEALs.

On this subject, David Fulghum wrote that the RQ-170 then carried a full-motion video (FMV) payload, noting that "FMV is the key to activity-based, intelligence analysis, the same discipline that revealed Osama bin Laden's hiding place. Both the CIA and the National Geospatial-Intelligence Agency (NGA) see activity-based intelligence as the path to better monitoring of areas of concern, and they are busy expanding that capability. . . . The single-channel, FMV capability is being multiplied up to 65 times in new systems being packaged for carriage by unmanned aircraft and airships. An Air Force version of the capability is Gorgon Stare. An Army system is called Argus-IS. . . . Gorgon Stare, developed by Sierra Nevada Corporation and the Air Force's Big Safari program, has been flying over Afghanistan on MQ-9 Reapers since December 2010. The current payload is in two pods. One carries a sensor ball produced by subcontractor ITT Defense. The ball contains five EO [electro-optical] cameras for daytime and four IR [infrared] cameras for nighttime ISR, positioned at different angles for maximum ground coverage. The pod also houses a computer processor. Images from the five EO cameras are stitched together by the computer to create an 80-megapixel image."

Both Sweetman and Fulghum were on the mark in suggesting that the Beast's primary mission was to snoop on places such as Iran. This intent became painfully clear on December 4, 2011, when one of the stealthy aircraft fell into the hands of the Iranian government near the city of Kashmar in northeastern Iran, 140 miles from the Afghan border.

Western news media reported that it had been "shot down," although when the Iranians put the RQ-170 on display it was clear that it had not been hit by a surface-to-air missile. The Iranians claimed that their cyber warfare experts took over the control telemetry channel and landed the aircraft. The US DOD said it was "flying a mission over western Afghanistan" when control was lost, adding that the aircraft had crash-landed.

Questions arose over whether the RQ-170 was being operated by the Defense Department, as initial comments suggested, or by the CIA. In the *Washington Post* of December 6, Greg Miller wrote that "CIA press officials declined to comment on the downed drone and reporters were directed toward a statement from the military. And sure enough, the NATO-led International Security Assistance Force seemed to step up to take the blame. 'The UAV to which the Iranians are referring may be a US unarmed reconnaissance aircraft that had been flying a mission over western Afghanistan late last week,' ISAF said in a statement. Mystery solved. The US military operates plenty of drones as part of the war effort in Afghanistan, and this one just veered off course. But the wording of the ISAF statement was curiously ambiguous, particularly on the question of who was really flying the drone. Some senior US officials seemed troubled by the attempt at deception from the start. On Sunday, a senior defense official voiced skepticism about the idea that a precious stealth drone would be doing surveillance work in western Afghanistan. By Monday, the story had changed. The CIA and the Pentagon continued to deny comment, but other US officials confirmed that the drone belonged to the CIA."

"Accurate information was provided in the statement," a "senior US official" told Miller. "There's no obligation to disclose all the details of sensitive reconnaissance missions. If that's the test, then we may as well knock on the doors of our adversaries, wherever they may be, and ask them to answer our questions."

With this, the United States predictably asked that the RQ-170 be returned, and the Iranians, just as predictably, refused. Also as could have been expected, the Iranians insisted that they were going to build a copy of the aircraft. As of February 2013, when Iran first released FMV footage downloaded from the RQ-170, the reverse-engineered replica had not yet appeared.

Back in the United States, a window into the future of twenty-first century black airplanes continued to be found in a look at industry-funded projects such as the Bird of Prey and the Polecat. Boeing and Lockheed Martin had not funded these programs internally on a whim, but under the assumption that the technology being developed would evolve into a salable future product—such as the RQ-170 and who knows what else.

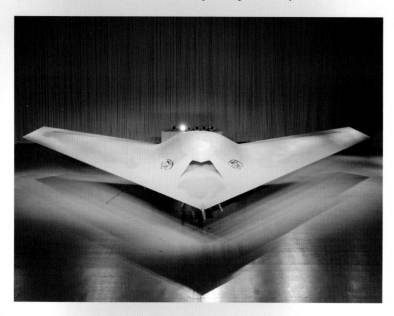

The Phantom Ray, the successor to the X-45C J-UCAS, made its first flight in 2011. *Chris Haddox, Boeing*

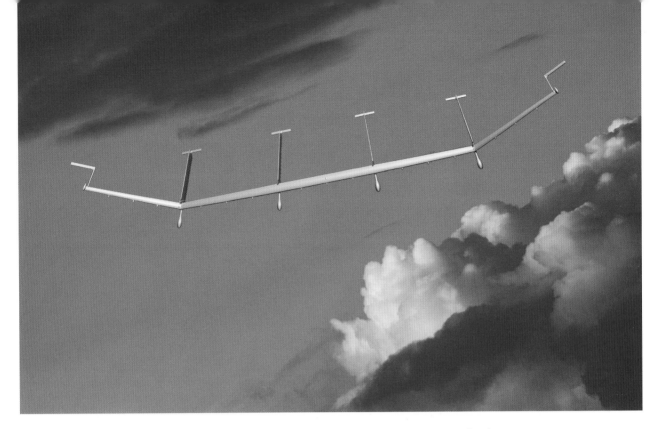

Another company-financed program is the Boeing Phantom Ray, named in part for the company's Phantom Works and in part for its shape. The Phantom Ray evolved from the X-45C J-UCAS, which was to have followed the X-45A UCAV but had been cancelled in 2006. In 2009, the same year that the Lockheed Martin RQ-170 was formally revealed, Boeing announced the decision to revive the X-45C as the company-funded Phantom Ray. "We will incorporate the latest technologies into the superb X-45C airframe design," said Dave Koopersmith, vice president of Boeing Advanced Military Aircraft, a division of Phantom Works. "Phantom Ray will pick up where the UCAS program left off in 2006 by further demonstrating Boeing's unmanned systems development capabilities in a fighter-sized, state-of-the-art aerospace system."

As Graham Warwick of *Aviation Week* wrote in May 2009, "If the [Phantom Ray] aircraft looks familiar, that's because it is—it's the X-45C that was completed, but never flown, when the [J-UCAS] program was cancelled back in 2006. . . . Unveiling of the Phantom Ray comes hard on the heels of US defense secretary Robert Gates' April 7 announcement that the [Next Generation Bomber] program is to be deferred and his comments that perhaps the next Air Force bomber could be unmanned. In effect, we are back to where we were before March 2006, when the J-UCAS program was planning to demonstrate technology for future unmanned strike/surveillance platforms."

The Phantom Ray entered its flight test program with a seventeen-minute first flight from Edwards AFB on April 27, 2011. While this first flight was announced, and it took place at an officially acknowledged facility rather than at Groom Lake, little more has been said officially, so it is hard to know the nature of the black world iceberg of which the Phantom Ray is the tip.

"The fact of the matter is that we have a stealthy, remotely piloted aircraft that's out there," General David Deptula, the Air Force Deputy Chief of Staff for Intelligence, Surveillance and Reconnaissance, told reporters back in 2010 when he was asked about the secrecy surrounding the RQ-170.

DARPA's Vulture was originally directed at developing the technology to allow a UAV to remain aloft for up to five years at a time. *DARPA*

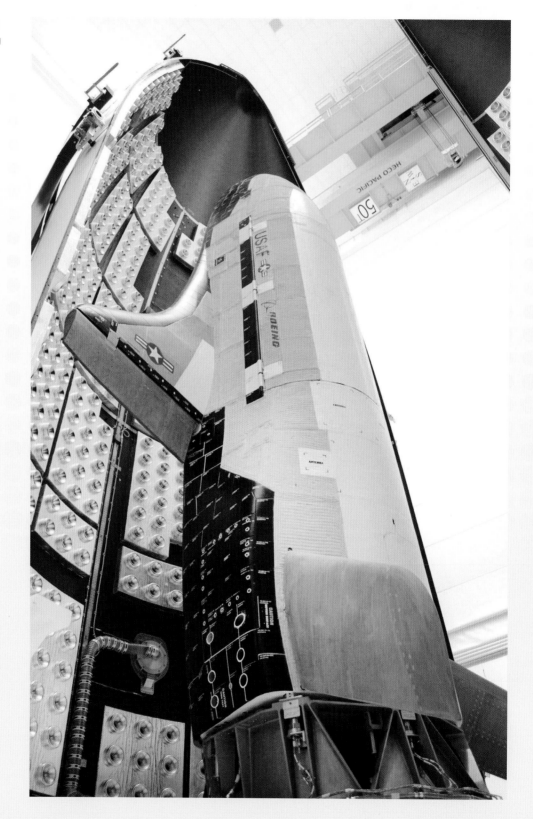

The X-37B Orbital Test Vehicle is shown being encapsulated for a March 2011 launch. In orbit for fifteen months, it came to earth for a runway landing. *USAF*

An X-51A Waverider is mounted under the wing of a B-52 for a Mach 4-plus test flight. *USAF*

He then went on to provide some insight into the next generation beyond the stealthy black world aircraft that now exist—or that are *known* to exist. "We can't do business in a serial fashion like we have before. We're not looking for the next version of the MQ-9 that can fly faster and go higher. Can we physically change the characteristics of an aircraft to adapt it to different roles by making it more survivable through shape and treatments?"

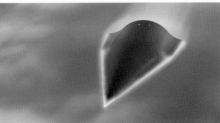

The Hypersonic Technology Vehicle is designed to fly at speeds up to Mach 20. *DARPA*

There are other ongoing, but little heralded, programs that provide tantalizing clues into what *might* be going on at Area 51 and the other areas about which we know even less. There is DARPA's Vulture program, aimed at developing a solar-powered UAV that can stay aloft for five years. The Boeing X-37B Orbital Test Vehicle is a UAV spaceplane that has already been operational in space on missions lasting more than a year.

Speed, as well as duration, has always been the hallmark of the black programs at Area 51. An indication of what might be going on in the black world is indicated by the Boeing X-51 Waverider. Powered by a Pratt & Whitney Rocketdyne SJY61 scramjet engine, it rides its own shockwave to hypersonic flight. By 2012, the Waverider had demonstrated speeds up to Mach 4.88.

The speed of the Waverider is eclipsed by that of Lockheed Martin's Hypersonic Technology Vehicle 2 (HTV-2), developed for a DARPA research and development effort known as Force Application and Launch from the Continental United States (FALCON). First test flown in 2011, it is, as DARPA explains, "an unmanned, rocket-launched, maneuverable aircraft that glides through the earth's atmosphere at incredibly fast speeds—Mach 20 (approximately 13,000 miles per hour). At HTV-2 speeds, flight time between New York City and Los Angeles would be less than 12 minutes."

The only certain thing about the skies over Area 51 are shapes and treatments that will keep black airplane speculators and aviation enthusiasts busy well into the twenty-first century.

EPILOGUE
STILL OUT THERE

AREA 51 IS LIKE a multifaceted gem. It is equal parts truth and illusion. It is equal parts hard, cold engineering reality and imaginative guesswork. It is equal parts playful obfuscation and "use-of-deadly-force" national security concealment and denial.

Area 51 is a black world of black airplanes that officially does not exist, and it is also a fantasy world of extraterrestrial visitation that almost certainly does not exist. It is the home of secret projects that did not exist—until we were told that they did exist. It is probably the home of secret projects that did exist, but about which we will never know.

It is a world of things that are perceived only by the shadows that they cast. As Bill Sweetman wrote in *Popular Science* magazine, the "vague, untraceable allocations in congressional budgets that often signal classified programs are on the rise, and modern technological innovations are now enabling aircraft designs that might have floundered in the black world for years. Further, there are significant gaps in the military's known aviation arsenal—gaps that the Pentagon can reasonably be assumed to be actively, if quietly, trying to fill."

Following the money into apparently benign voids is like trying to follow a magician's sleight of hand or a Las Vegas blackjack dealer's practiced hand.

Is it any wonder that the gateway to the mystery world of Area 51 should be an airport in the capital of fantasy? Can the boundary between fact and illusion be fuzzier anywhere on earth than it is in Las Vegas—*except* perhaps, at Area 51?

People who fly in and out of Las Vegas hardly notice a nondescript terminal off at the northwest corner of the 2,800-acre sprawl of McCarran International Airport. Nor does the majority notice the white Boeing 737s that are parked at this nondescript terminal.

With the lights of the celebrated "Strip" barely a half mile away, why should they? These visitors have come fixated not on unmarked airport buildings but on the well-delineated pleasures of this unique and sparkling, larger-than-life theme park. As they deplane, anxious for a weekend or more of bachelor parties, of nightclubs, of luxuriating their senses, or of gaming, they hardly notice as one of those plain white 737s taxis out, takes to the air and heads due north into the darkening sky and into an unknown world that is perpetually "dark" in the metaphorical way.

A crewed hypersonic aircraft, the once and future image of the aircraft from Area 51. Illustration by Erik Simonsen

"Janet" is the call sign used by this fleet of airplanes operating from Las Vegas to Groom Lake and Tonopah. The Janet fleet, fewer than a dozen in number, include Boeing 737s and Beechcraft executive propliners. They are said to be owned by the US Air Force, but they carry civilian registration. They are painted white and carry no markings other than a tail number and a red line on each side of the fuselage.

Janet is the tangible, yet mysterious, link to that other world. She is the means by which civilian and military personnel working in that world pass through the looking glass, and they are our tangible indication that the looking glass has another side.

In the early days of Groom Lake operations, when most of the civilians traveling in and out were Lockheed employees working on the Aquatone and Oxcart programs, the US

Air Force operated routine flights between Burbank and Area 51. According to Gregory Pedlow and Donald Welzenbach, "The project staff decided that the simplest approach would be to fly the essential personnel to the site on Monday morning and return them to Burbank on Friday evening. Frequent flights were also necessary to bring in supplies and visitors from contractors and headquarters."

In those days, transportation was provided by a regularly scheduled Military Air Transport Service flight using a C-54 that was known as "Bissell's Narrow-Gauge Airline" after CIA overhead reconnaissance programs chief Richard Bissell.

By the 1970s, much of the site management at Groom Lake, like that at the neighboring NTTR, was outsourced to civilian contractors. One of these was the engineering firm of Edgerton, Germeshausen, and Grier (EG&G), which first entered the world of nuclear testing and black airplanes to develop systems to monitor and evaluate experimental technologies. Gradually, EG&G took on a wider facilities management role at secure government locations, and they have played a key role at Groom Lake and Tonopah.

EG&G has also been a prominent part of the Area 51 conspiracy theory lore for decades. A mere mention of their name will elicit a knowing nod from any black airplane enthusiast. They are believed to be the operators of the Janet airline and to have the contract for guarding the perimeter of the Groom Lake complex.

Since 2002, EG&G has been fully absorbed into another engineering and management firm, URS Corporation (formerly United Research Services). Among its white world activities, the EG&G Division of URS entered into an "institutional services contract" to help manage NASA's Kennedy Space Center.

Area 51's perimeter is still a tourist attraction for UFO and black airplane enthusiasts. The official Highway 375 sign says "Extraterrestrial," but the picture is of an F-117A. *Bill Yenne*

Traveling to Groom Lake from Las Vegas via Janet takes less than an hour. For the majority of us who cannot book a flight on this successor to Bissell's airline, the drive takes hours—from the wall-to-wall-crowd density of Las Vegas through some of the emptiest country in the contiguous United States. To actually see what it is all about amid the mysterious mountains of the Nellis Range, this long drive is compulsory.

A "Janet" 737 rests at its unmarked terminal, close to the Las Vegas Strip. *Bill Yenne*

The runway and hangar complex at Groom Lake officially did not exist when these pictures were taken in 1998. Fifteen years later, the US Air Force inadvertently released the photo seen on pages 10 and 11 of this book, and the lid of secrecy was edged back. *Tony Landis photos*

The exit off Interstate 15 for Nellis AFB is only eighteen miles from McCarran, and just a dozen miles later, as you turn north onto US Highway 93, there is nothing ahead or in the rear-view mirror but desert. The ninety-three miles of two-lane Highway 93 that run north to the junction of Nevada Highway 375, the Extraterrestrial Highway, are long and straight. It is the kind of road where 80 mph feels like 50 mph. It is populated by long haul truckers, the occasional motor home, a few ranchers in their pickups, and the handful of people who are drawn to this quarter by the mystery stories told on Internet sites or whispered in faraway coffee shops.

Halfway to Ash Springs, a sign warning of low flying aircraft marks the point at which Highway 93 passes beneath the edge of the air space controlled by Nellis AFB as part of the Desert Military Operating Area (MOA). For the remainder of the drive, nothing in the sky above is there without permission of the Nellis Tower. During Red Flag, the skies here are filled with F-15s, B-2s, and all manner of hardware, but the mysterious airspace of Area 51 is still far ahead.

There is a welcome gas station in Ash Springs, but little else can be seen along Highway 93 other than the primordial desert landscape and the intriguing contrails high above. Turning off at the junction just north of Ash Springs, one finds that the emptiness of the Extraterrestrial Highway makes Highway 93 seem like the Las Vegas Strip. It is one of those highways where three or four songs can go by on your music player before you pass another car. Mostly, it is local traffic, but you can tell by the bumper stickers on some of the vehicles that people are still coming out here to squint at those contrails and to look longingly for lights in the night sky.

There are no gas stations on the Extraterrestrial Highway or fences to keep wandering cattle off the road. The landscape is unchanged since the nineteenth century settlers passed this way and decided *not* to stay.

Passing over the Pahranagat Range at Hancock Summit the traveler is greeted with a view of a distant desert landscape scarred by the longest, straightest gravel road imaginable. Some have compared it to the Nazca Lines in Peru, which Erich von Däniken famously postulated to have been made by extraterrestrials. This one was not made by extraterrestrials, but many people believe that it *leads* to them.

This line is the road into Groom Lake, to Area 51. It was built before von Däniken's best-selling book *Chariots of the Gods?* was first published in 1968 and two decades before Bob Lazar drove this road on his way to work. Beyond the point where the road disappears into the horizon, one can see the ridge that obscures the view of Groom Lake. From near Hancock Summit, those with four-wheel-drive vehicles—or a blatant disregard for rental cars—can drive most of the way to a place on Tikaboo Peak where the actual facilities at Groom Lake can be viewed at a distance of about forty miles.

The turn-off to the Groom Lake Road at Lincoln County Milepost 34.6 is obvious but unmarked. Nearby, at Milepost 29.6, is one of the principal icons of Area 51 folklore. Though it has been painted white since around 1996, it is still known as the Black Mailbox.

The mailbox actually belongs to Steve Medlin, whose ranch is nearby, but some conspiracy theorists believe this to be a mere cover story. Despite the fact that it bears Medlin's name and contains his mail, rumors still persist that this is the place where top secret mail is delivered to anyone and everyone from the Men in Black of 1950s flying saucer mythology to the extraterrestrials themselves. Today, it also serves as a message board for the enthusiasts who cannot resist the urge for graffiti.

The road that leads from the Black Mailbox past Medlin's ranch intersects the Groom Lake Road. This modern Nazca Line is one of the best-maintained gravel roads anywhere. As you drive, it is easy to find your mind wandering back to February 26, 1962, when Article 121, the first Oxcart A-12, was trucked to Groom Lake over this same road. If the nearby Joshua trees could talk, what stories they could tell of the exotic hardware whose conveyances have kicked up dust here.

The trees, except in the most outlandish fantasies, do not have voices, but it is widely believed by those who spend their days in Area 51 speculation, that the road has "ears"—sensors buried along the way to alert the guards of incoming visitors. As anyone can see, the hills *do* have eyes. No attempt is made to disguise the camera stands on nearby Bald Mountain.

After nearly fourteen miles on the straight section, Groom Lake Road rounds a corner and reaches the border of the restricted zone. There is no gate and no fence, but the unambiguous warning signs leave no doubt that *this* is the point beyond which outsiders dare not go.

During the Cold War, AFB perimeters were marked with signs that carried the warning "Use of Deadly Force Authorized," meaning that the guards were permitted to *kill* trespassers. These are gone now, but the threat of detainment and arrest lives on. Not so high up on the hills nearby, watching anyone who stops to view the signs and make the obligatory U-turn, are guards in pickups and Jeep Cherokees watching with binoculars.

As their vehicles are not in military colors, it is assumed that these people are employees of a contractor firm such as EG&G or URS. Area 51 enthusiasts call them the "Cammo Dudes," because they do wear camouflaged uniforms. This is counterintuitive, because their primary function, short of making arrests, is to *be seen*.

Located about twenty miles northwest of the Black Mailbox, Rachel, Nevada, is the only town on the Extraterrestrial Highway. An icon of the Area 51 subculture in its own right, Rachel experienced its heyday in the 1990s during the decade after the official revelation of the F-117A and the unofficial revelations of Bob Lazar.

Below left: How can photography of something that officially does not exist be prohibited? *Bill Yenne*

Below right: This is the end of the line on the Groom Lake Road. The men in the vehicle atop the hill will ensure that the warnings are enforced. *Bill Yenne*

Both Aliens and Earthlings are welcome here at the Little A'Le'Inn. So, too, are people who work at the "nonexistent" Groom Lake facility. *Bill Yenne*

During those days, Rachel was *the* place to visit for people yearning to look at lights in the sky, as well as those yearning to look at the people who believed in lights in the sky. Conspiracy theorists and media stars came out to the desert to peer across the hills. In April 1994, ABC Television's *World News Tonight*, hosted by Peter Jennings, sent a crew to Rachel to interview the local conspiracy theorists and to attempt to peek into Area 51. Five months later, talk show legend Larry King came to Nevada to do a show on Rachel and Area 51.

Twentieth Century Fox and director Roland Emmerich visited Rachel and environs to film the signature scenes for the 1996 alien invasion blockbuster, *Independence Day*. When the United States government refused to cooperate with the filming because the studio refused to delete mentions of Area 51, the entire cache of the conspiracy mythology grew even larger. The film went on to win a Best Special Effects Oscar, and was, for a while, one of the two top-grossing films of all time. The producers donated a large concrete time capsule to Rachel, which is still there.

If Rachel in the 1990s had been a movie, the lead character would have been a computer programmer from Boston named Glenn Campbell, who dropped out and moved here in 1993. Calling himself "PsychoSpy," he became the one-man central clearing house for all UFO and Area 51 information and established the Area 51 Information Center in a mobile home. He led tours to a hilltop he dubbed "Freedom Ridge," which overlooked Groom Lake. For two years, he fought a losing battle with the US Air Force, who eventually cut off public access to this vantage point in 1995.

From 1994 to 1996, Campbell published a newsletter called *The Groom Lake Desert Rat*, which offered a tantalizing insight into the often competing worlds of UFOs and black airplane enthusiasts. In his 1998 book, *Dreamland*, Phil Patton variously describes Campbell as "the park ranger of Dreamland," a "jester," a "philosopher," and a "naturalist of the unnatural" who sometimes "suggested a parodic Thoreau, with Groom Dry Lake as his Walden Pond."

Patton goes on to write that "if he figured as the Hamlet of the hamlet of Rachel, Campbell was also the village explainer, laying out the mythologies, systematizing the lore. Glenn Campbell was the closest thing in Rachel to Joseph Campbell. In his role as PsychoSpy, he was drawn to the tales as parables. In one tale, he was lying in the back seat of his car parked along Mailbox Road when he first saw them: strange spaceships, dotted with lights, hovering. They flew right over the car. It was only later, after thinking about the vivid memories he had had, that he realized he had been lying in a position from which he could not have seen ships overhead. He used the story as an example of how easy it was to delude yourself into thinking you had seen something you had not, how tricky the business of seeing things in the sky near the Black Mailbox was."

In those days, the Rachel city limits sign carried a population reckoning that read "Humans 98, Aliens ??" The official 2010 Census counted 54 humans but did not mention aliens.

Since the turn of the century, Rachel has quieted down. In 2001, Glenn Campbell shuttered his Area 51 Information Center and moved on. The town's only gas station closed in 2006. If a traveler failed to get gas in Ash Springs, the next option is in Tonopah, 110 miles away.

However, there is one institution that existed in Rachel before Glenn Campbell came west and has survived his departure by more than a decade. Back in 1989, Pat Travis and her husband Joe took over the Rachel Bar and Grill and renamed it the Little A'le'Inn. They started selling UFO collectibles and military shoulder patches and serving alien burgers and beer.

"We have entertained visitors from all over the world," Pat explains. "We love to meet new people and make new life friends as we have done over the many years."

In small-town America, the Main Street coffee shop is the place where world affairs are discussed and decided. In Rachel, where the Extraterrestrial Highway *is* Main Street, the Little A'le'Inn is the forum of discussion. It is a place where you order a hamburger, and they ask whether you want an *Alien* Burger in the same way fast food places used to ask whether you wanted your order to be "super-sized."

Over a burger and a cup of earthly coffee, patrons get down to business.

On my last trip, some tourists were asking about the Black Mailbox. I was pleased with myself for recalling the milepost number. In unison, the waitress and I added, "It's painted white now."

At the Little A'le'Inn, everyone has a story, and most of the regulars have seen "something," whether it be lights in the sky or odd things flying by daylight. One recalls a nearby crash a number of years ago but adds that it was "an Air Force jet, not a UFO." The pilot got out okay.

At Rachel, we are under the skies of the Desert MOA, so *anything* that is up there is different from what most of us see back home. The regulars explain that every day people

"I Want 2 Believe," reads a poignant message scrawled on the Black Mailbox. Long ago painted white, it is the essential shrine of UFO enthusiasts on the Extraterrestrial Highway. Note the persistent lenticular clouds over Area 51, which lies just across the Groom Mountains in the distance. Today, the mailbox, actually the real mailbox of rancher Steve Medlin, also serves as an important bulletin board for travelers who pass this way in search of aliens, black airplanes, and other mysteries in which they want to believe.
Bill Yenne

from far away come here to see those things—or at least to be able to say that they looked. Some are passionate to the extreme about what they hope to see. The locals describe some of these enthusiasts as being stranger than the aliens.

The locals tell us that more Canadians and Britons than Americans stop in at the Inn: a group from France, a bus full of tourists from Japan who were taking pictures of everything. They meet all kinds. They meet people who tell of conversations overheard while listening to police scanners out here. One zealous believer insisted that the fact of a zillion stars in the galaxy is proof positive that they have extraterrestrials at Area 51. Another recounted stories told online and is answered by a cynic who reminds her in an ironic tone that "if it's on the Internet, it *must* be true."

The conversation is more about black airplanes than captured extraterrestrials, but mainly it is about the *place*. People are curious about the four-wheel-drive road up Tikaboo Peak and about what happens to people who are caught venturing across the imaginary line guarded by the Cammo Dudes. The general consensus, shared by tourists and locals alike, is that Area 51 is still a hive of activity. The watchers of Google Earth report facilities still under construction at Groom Lake, and *Aviation Week* continues to report sightings, both out here and in the deliberately obscure line items in the federal budget.

The road into the "Back Gate" of Area 51 departs the Extraterrestrial Highway just a short distance south of Rachel, so people who work on the other side of the fence also occasionally stop by the Little A'le'Inn for a burger and a cup of coffee. When asked what they do "in there," they seem invariably to describe themselves as "janitors." EG&G *is*, after all, a site management company.

Nobody will admit to having actually *seen* an alien in the Little A'le'Inn. We are all strangers here, so how would anyone know?

Stepping out into the lengthening shadows of a late December afternoon, we notice lenticular clouds dotting the sky, and excitement laces the air. It is the excitement of being at the edge of the unknown and being able to sense its secrets.

Driving south along the Extraterrestrial Highway at the edge of night, we pause in the cold quiet of a nearly empty highway to ponder the mysteries of this place.

In the skies over Area 51, there *are* lights.

Are they the lights of the routine Janet or of something else? There are lights that seem to pause and hover, then scoot away in another direction. Was it a 737 making a turn or something else?

There is no doubt.

They *are* testing airplanes on the other side of that ridge.

These airplanes *will* amaze and delight us—when and *if* we are told about them.

There are things out there about which we have never heard but *hope* we will.

These things will possess features that will be as amazing as the speed of Oxcart was in the 1960s or as the stealthiness of Have Blue was in the 1980s.

Perhaps they will possess features or deliver performance that we had never imagined.

All we know is that they *are* out there.

In reflecting on the decades since humans first brought airplanes to Groom Lake, we hear the voices of Kelly Johnson, of Ben Rich, and of countless others who made history here. Just as Bill Sweetman studies government paperwork, searching for the voids that tell the shape of the unknown, we think of the voices that we will never hear, and we yearn to hear the tales that *could* be told about the unknown—but will never be told.

Finally, with the vastness of the Nellis Range and the sky above it spread out before us, we recall the advice cordially given by Pat at the Little A'le'Inn when she explained that we should "look up, as the truth lies there. Always keep your eyes to the skies whenever you can. You just never know when that special event will happen. At those times there may be no answers, leaving you only to wonder what just happened or what you saw, and having to ask more questions . . . getting no answers. Life is a mystery, enjoy the ride. The events and unidentified flying objects we see and only hear at times in this area often leave us shaking our heads. The unknown is what we live for. The times when logic escapes us and the knowledge of things to come are before us. We welcome it all."

That truly sums it up.

This slightly retouched photo was taken near twilight on Highway 93 beneath the restricted air space of the Nellis Range, about forty-two miles east of Groom Lake. *Bill Yenne*

ACRONYMS

AAF	Army Airfield
AB	Air Base
ACM	Air Combat Maneuvers (or Air Combat Maneuvering)
ACM	Advanced Cruise Missile
ADP	[Lockheed] Advanced Development Projects (the "Skunk Works")
AEC	Atomic Energy Commission
AFB	Air Force Base (US Air Force)
AFS	Air Force Station (US Air Force)
AFSC	Air Force Systems Command (US Air Force)
AFSWP	Armed Forces Special Weapons Project
AMARC	Aerospace Maintenance and Regeneration Center
AMC	Air Materiel Command (US Air Force)
APG	Aerial Phenomena Group
ARDC	Air Research and Development Command (US Air Force)
ARS	Aerial Refueling Squadrons
ATA	Advanced Tactical Aircraft
ATB	Advanced Technology Bomber
ATC	Air Training Command
ATC	Air Training Command (US Air Force)
ATF	Advanced Tactical Fighter
ATIC	Air Technical Intelligence Center (US Air Force)
AWACS	Airborne Warning and Control System
BSAX	Battlefield Surveillance Aircraft, Experimental
BVR	Beyond Visual Range
C3	Command, Control, and Communications
CIA	Central Intelligence Agency
DARO	Defense Airborne Reconnaissance Office
DARPA	Defense Advanced Research Projects Agency
DCI	Director of Central Intelligence
DDCI	Deputy Director of Central Intelligence
DMZ	Demilitarized Zone
DOD	US Department of Defense
DPD	Development Projects Division
EG&G	Edgerton, Germeshausen, and Grier

FAA	Federal Aviation Administration
FMV	Full–Motion Video
FTD	Foreign Technology Division
FWS	Fighter Weapons School
GAO	General Accounting Office
HAE	High Altitude Endurance
HALE	High-Altitude, Long-Endurance
HTV-2	Hypersonic Technology Vehicle 2
ICBM	Intercontinental Ballistic Missile
JRC	Joint Reconnaissance Center
J-UCAS	Joint Unmanned Combat Air System
JSTAR	Joint Surveillance Target Attack Radar System
KGB	Komitet Gosudarstvennoy Bezopasnosti (Soviet Committee for State Security)
LASRE	Linear Aerospike SR-71 Experiment
LO	Low Observable
MASDC	Military Aircraft Storage and Disposition Center
MATS	Military Air Transport Service (US Air Force)
MOA	Military Operating Area
MRBM	Medium-Range ballistic missile
N2S2	Nevada National Security Site (formerly NTS)
NACA	National Advisory Committee for Aeronautics (became NASA in 1958)
NASA	National Aeronautics and Space Administration
NASM	National Air & Space Museum
NAVAIR	Naval Air Systems Command
NDIA	National Defense Industrial Association
NEPA	Nuclear Energy for Propulsion of Aircraft
NERVA	Nuclear Engine for Rocket Vehicle Application
NRO	National Reconnaissance Office
NSC	National Security Council
NTS	Nevada Test Site (later N2S2)
NTTR	Nevada Test and Training Range
OSA	Office of Special Activities
OSS	Office of Strategic Services (World War II)
PTD	Physical Theory of Diffraction

RCS	Radar Cross Section		TTR	Tonopah Test Range
RIFT	Reactor In Flight Test		UAS	Unmanned Aerial System
ROCAF	Republic of China Air Force		UAV	Unmanned Aerial Vehicle
RPV	Remotely Powered Vehicle		UCAS	Unmanned Combat Air System
RSO	Reconnaissance Systems Officer		UCAV	Unmanned Combat Air Vehicle
SAB	Scientific Advisory Board (US Air Force)		UFO	Unidentified Flying Object
SAC	Strategic Air Command (US Air Force)		USAAF	United States Army Air Force
SAM	Surface-to-Air Missile		USAFE	United States Air Forces in Europe
SAS	Special Activities Squadron		VTOL	Vertical Take-Off and Landing
SRS	Strategic Reconnaissance Squadrons		XECF	Experimental Engine Cold Flow
SRW	Strategic Reconnaissance Wing		XST	Experimental Survivable Testbed
TFTS	Tactical Fighter Training Squadron			
TFS	Tactical Fighter Squadron			
TFW	Tactical Fighter Wing			
TMC	Titanium Metals Corporation			

ABOUT THE AUTHOR

Bill Yenne is the author of more than three dozen nonfiction books, especially on aviation and military history. These have included profiles of the B-52 Stratofortress, unmanned combat air vehicles, and secret weapons of the Cold War, as well as histories of the Strategic Air Command, the US Air Force, and his recently updated *The Story of the Boeing Company*. He has contributed to encyclopedias of both World War I and World War II and has appeared in television documentaries on the History Channel, the National Geographic Channel, and ARD German Television. He visited the U-2 and SR-71 in their heyday at their rookery at Beale AFB, and he has traveled the lonely desert perimeter of Area 51.

Mr. Yenne is also a member of the American Aviation Historical Society. He lives in San Francisco and on the worldwide web at www.BillYenne.com.

BIBLIOGRAPHY

Air Technical Intelligence Center (ATIC). *Technical Report No. 102-AC 49/15-100 (Project Grudge)*. Wright Patterson AFB, Ohio: Air Materiel Command, 1949.

Bechloss, Michael. *May-Day: Eisenhower, Khrushchev and the U-2 Affair*. New York: Harper & Row, 1986.

Bissell, Richard M. Jr., with Jonathan E. Lewis and Frances T. Pudlo. *Reflections of a Cold Warrior: From Yalta to the Bay of Pigs*. New Haven and London: Yale University Press, 1996.

Brandt, Steven A., Randall J. Stiles, and John J. Bertin. *Introduction to Aeronautics: A Design Perspective*. Reston, Virginia: American Institute of Aeronautics & Astronautics, 2004.

Brown, Kevin V. "America's SuperSecret Spy Plane," *Popular Mechanics*, June 1968.

Brown, William H. "J58/SR-71 Propulsion Integration," *Studies in Intelligence*, Summer 1982.

Condon, Dr. Edward U., et. al. *Scientific Study of Unidentified Flying Objects (SSUFO) undertaken under Air Force Scientific Advisory Board (SAB) Contract F44620-67-C-0035*. Boulder: The University of Colorado, 1969.

Crickmore, Paul, and Jim Laurier. *Lockheed SR-71 Operations in the Far East*. Oxford, UK: Osprey Publishing, 2008.

Crickmore, Paul F. "Blackbirds in the Cold War." In *Air International*. Stamford, UK: Key Publishing, 2009.

Crickmore, Paul F. "Lockheed's Blackbirds – A-12, YF-12 and SR-71A." In *Wings of Fame*. London: Aerospace Publishing, 1997.

Crickmore, Paul F., and Alison J. *Nighthawk F-117 Stealth Fighter*. St. Paul, Minnesota: Motorbooks, 2003.

Crickmore, Paul F. *Lockheed Blackbird: Beyond the Secret Missions*. Oxford, UK: Osprey Publishing, 2004.

Crickmore, Paul F. *Lockheed SR-71: The Secret Missions Exposed*. London: Osprey, 1996.

Davies, Steve. *Red Eagles: America's Secret MiGs*. Oxford, United Kingdom: Osprey, 2008.

Donald, David, ed. "Lockheed's Blackbirds: A-12, YF-12 and SR-71." *Black Jets*. Norwalk, CT: AIRtime, 2003.

Dorr, Robert. "F-117: The Black Jet," *World Air Power Journal*. London: Aerospace Publishing, 1994.

Drendel, Lou. *SR-71 Blackbird In Action*. Carrollton, TX: Squadron/Signal Publications, 1982.

Durham, H. B. *Raypac: A Special-Purpose Analog Computer, Report Numbers AECU-3213; TM-46-55-54*. Albuquerque: Sandia Corporation, 1955.

Fehner, Terrance R., and F. G. Gosling. *Origins of the Nevada Test Site (DOE/MA-0518)*. Washington, DC: History Division, US Department of Energy, 2000.

Goodall, James, and Jay Miller. *Lockheed's SR-71 "Blackbird" Family: A-12, F-12, M-21, D-21, SR-71*. Hinckley, UK: Midland Publishing, 2002.

Goodall, James. *SR-71 Blackbird*. Carrollton, TX: Squadron/Signal Publications, 1995.

Graham, Richard H. *SR-71 Blackbird: Stories, Tales, and Legends*. St. Paul, Minnesota: Zenith Press, 2002.

Graham, Richard H. *SR-71 Revealed: The Inside Story*. St. Paul, Minnesota: MBI Publishing Company, 1996.

Grant, R. G. *Flight: 100 Years of Aviation*. New York: DK Publishing, 2007.

Hacker, Barton C. *Elements of Controversy: The Atomic Energy Commission and Radiation Safety in Nuclear Weapons Testing, 1947-1974*. Berkeley: University of California Press, 1994.

Haines, Gerald K. "The CIA's Role in the Study of UFOs, 1947-90," *Studies in Intelligence* 41:1, 1997.

Helms, Richard, with William Hood. *A Look Over My Shoulder: A Life in the Central Intelligence Agency*. New York: Random House, 2003.

Hobson, Chris. *Vietnam Air Losses, USAF, USN, USMC, Fixed-Wing Aircraft Losses in Southeast Asia 1961-1973*. North Branch, Minnesota: Specialty Press, 2001.

Jenkins, Dennis R. *Lockheed Secret Projects: Inside the Skunk Works*. St. Paul, Minnesota: MBI Publishing Company, 2001.

Jenkins, Dennis R. *Lockheed SR-71/YF-12 Blackbirds*. North Branch, Minnesota: Specialty Press, 1997.

Johnson, Clarence L. "Kelly," with Maggie Smith. *More Than My Share Of It All*. Washington, DC: Smithsonian Institution Press, 1985.

Johnson, Clarence L. "Kelly." *Kelly Johnson Papers, "Log for Project X."* Burbank, California: Lockheed Corporation, Advanced Development Projects Division, 1954–1955.

Johnson, Clarence L. "Kelly." "Development of the Lockheed SR-71 Blackbird," *Studies in Intelligence*, Summer 1982.

Khrushchev, Sergei N. *Nikita Khrushchev and the Creation of a Superpower*. State College, PA: Penn State Press, 2000.

Land, Edwin. *Memorandum for DCI Allen F. Dulles: A Unique Opportunity for Comprehensive Intelligence*, November 5, 1954.

Landis, Tony R., and Dennis R. Jenkins. *Lockheed Blackbirds*. Minneapolis, Minnesota: Specialty Press, revised edition, 2005.

Landis, Tony R. *Lockheed Blackbird Family: A-12, YF-12, D-21/M-21 and SR-71 Photo Scrapbook*. North Branch, Minnesota: Specialty Press, 2010.

Manning, Thomas A. *History of Air Education and Training Command, 1942–2002*. Randolph AFB, Texas: Office of History and Research, Headquarters, AETC, 2005.

Masters, David. *German Jet Genesis*. London: Jane's, 1982.

McIninch, Thomas. *The Oxcart Story*. Langley, Virginia: Center for the Study of Intelligence, Central Intelligence Agency, 1996.

Merlin, Peter W. "The Truth is Out There. SR-71 Serials and Designations." In *Air Enthusiast*. Stamford, UK: Key Publishing, 2005.

Merlin, Peter W. *From Archangel to Senior Crown: Design and Development of the Blackbird*. Reston, Virginia: American Institute of Aeronautics and Astronautics (AIAA), 2008.

Merlin, Peter. *Mach 3+: NASA USAF YF-12 Flight Research 1969-1979*. Washington, D.C.: Diane Publishing Co., NASA History Division Office, 2002.

Military Division of the Supreme Court of the U.S.S.R. *The Trial of the U-2: Exclusive Authorized Account of the Court Proceedings of the Case of Francis Gary Powers, Heard before the Military Division of the Supreme Court of the U.S.S.R., Moscow, August 17, 18, 19, 1960*. Chicago: Translation, World Publishers, 1960.

Miller, Jay. *Lockheed's Skunk Works*. Arlington, TX: Aerofax, 1993.

Miller, Jay. *The X-Planes: X-1 to X-45*. Hinckley, UK: Midland, 2001.

National Security and International Affairs Division. *Operation Desert Storm: Evaluation of the Air Campaign (GAO/NSIAD-97-134)*. Washington, DC: National Security and International Affairs Division (NSIAD), General Accounting Office (GAO), 1997.

Pace, Steve. *Lockheed SR-71 Blackbird*. Swindon, UK: Crowood Press, 2004.

Pappas, Terry. "The Blackbird is Back." *Popular Mechanics*, June 1991.

Patton, Phil. *Dreamland: Travels Inside the Secret World of Roswell and Area 51*. Villard/Random House, 1998.

Pedlow, Gregory W., and Donald E. Welzenbach. *The Central Intelligence Agency and Overhead Reconnaissance: The U2 and OXCART Programs, 1954-1974*. Langley, Virginia: Central Intelligence Agency, 1992.

Pocock, Chris. "U-2: The Second Generation." *World Air Power Journal*. London: Aerospace Publishing, 1997.

Powers, Francis Gary, and Curt Gentry. *Operation Overflight*. London: Hodder & Stoughton Ltd, 1971 (hard cover) Potomac Book, 2002 (paperback).

Reithmaier, Lawrence W. *Mach 1 and Beyond*. New York: McGraw-Hill, 1994.

Remak, Jeannette, and Joseph Ventolo, Jr. *A-12 Blackbird Declassified*. St. Paul, Minnesota: MBI Publishing Co., 2001.

Remak, Jeannette, and Joseph Ventolo, Jr. *The Archangel and the OXCART: The Lockheed A-12 Blackbirds and the Dawning of Mach III Reconnaissance*. Bloomington, IN: Trafford Publishing, Co., 2008.

Rich, Ben R., and Leo Janos. *Skunk Works: A Personal Memoir of My Years at Lockheed*. Boston: Little, Brown and Company, 1994.

Richelson, Jeffrey T. *The Wizards of Langley: Inside the CIA's Directorate of Science and Technology*. Boulder, CO: Westview Press, 2001.

Rose, Bill. *Secret Projects: Military Space Technology*. Hinckley, England: Midland Publishing, 2008.

Ruppelt, Edward J. *The Report on Unidentified Flying Objects*. Garden City, New York: Doubleday & Company, 1956.

Shul, Brian, and Sheila Kathleen O'Grady. *Sled Driver: Flying the World's Fastest Jet*. Marysville, California: Gallery One, 1994.

Stevenson, James P. *The $5 Billion Misunderstanding: the Collapse of the Navy's A-12 Stealth Bomber Program*. Annapolis, Maryland: Naval Institute Press, 2001.

Suhler, Paul A. *From Rainbow to GUSTO: Stealth and the Design of the Lockheed Blackbird*. Reston, VA: American Institute of Aeronautics and Astronautics, 2009.

Sweetman, Bill. "Aurora – is Mach 5 a reality?" *Interavia Aerospace Review*, November 1990.

Sweetman, Bill. *Aurora: The Pentagon's Secret Hypersonic Spyplane*. St. Paul, Minnesota: Motorbooks International, 1993.

Sweetman, Bill. *Lockheed Stealth*. St. Paul, Minnesota: MBI Publishing, 2001.

Ufimtsev, Pyotr Yakovlevich. *Fundamentals of the Physical Theory of Diffraction*. Hoboken, New Jersey: Wiley & Sons, Inc., 2007.

Ufimtsev, Pyotr Yakovlevich. *Method of Edge Waves in the Physical Theory of Diffraction*. Moscow: Soviet Radio, 1962.

Ufimtsev, Pyotr Yakovlevich. *Theory of Edge Diffraction in Electromagnetics*. Encino, California: Tech Science Press, 2003.

von Däniken, Erich. *Chariots of the Gods? Unsolved Mysteries of the Past*. New York: Putnam, 1968.

West, Nigel. *Seven Spies Who Changed the World*. London: Secker & Warburg, 1991 (hard cover). London: Mandarin, 1992.

Wheelon, Albert D. "And the Truth Shall Keep You Free: Recollections by the First Deputy Director of Science and Technology," *Studies in Intelligence*, Spring 1995.

Whittenbury, John R. "From Archangel to OXCART: Design Evolution of the Lockheed A-12, First of the Blackbirds." CIA PowerPoint presentation, August 2007.

Yenne, Bill. "Chapter 10: Stealth Aircraft." In *Secret Weapons of the Cold War: From the H-Bomb to SDI*. New York: Berkley Publishing Group, 2005.

Yenne, Bill. *B-52: The Complete History of the World's Longest Serving and Best-Known Bomber*. Minneapolis, Minnesota: Zenith Press, 2012.

Yenne, Bill. *Lockheed*. New York: Random House, 1987.

Yenne, Bill. *The History of the US Air Force*. New York: Simon & Schuster, 1984, 1992.

Yenne, Bill. *The Story of the Boeing Company*. San Francisco/Minneapolis: AGS BookWorks/Zenith Press, 2005, 2010.

INDEX